Kent Electronics Project

ASHLEY ROSS
TEACHER/ADVISER FOR ELECTRONICS

MAX HORSEY
HEAD OF SCIENCE, WARREN WOOD SECONDARY SCHOOL FOR BOYS

PETER HOOKER
SCIENCE INSPECTOR

TEACHER'S GUIDE

Oliver & Boyd

Cover artwork by Trevor Carter
Diagrams by John Marshall
Photographs by Andrew Allan

Oliver & Boyd
Robert Stevenson House
1–3 Baxter's Place
Leith Walk
Edinburgh EH1 3BB

A Division of Longman Group UK Ltd

ISBN 0 05 004285 8

First published 1988

Set in 11 on 13pt Linotron Palatino.

Produced by Longman Group (FE) Ltd.
Printed in Hong Kong.

CONTENTS

THE KENT ELECTRONICS KIT

The **Kent Electronics Kit**, containing 18 circuit boards plus plug-in connectors, is available from Griffin and George, Bishop Meadow Road, Loughborough, Leicestershire LE11 ORG. A supplementary kit of Nuffield Extension Boards is also available for Investigations 42 to 48.

ELECTRONICS IN GCSE

The importance of electronics as an area of study in the school curriculum is now well recognised. Most examination boards have included an element of electronics within GCSE physics, whilst within CDT electronics can be used most effectively as a problem-solving mechanism. Often, a module of electronics is to be found in modular science courses, as well as in integrated and coordinated science schemes. With science and technology taking up a substantial part of the curriculum between them, it seems inevitable that electronics will have a major role to play.

In a society which has become increasingly dependent on technology, it would be difficult to imagine going through a day without the help of electronic devices. Indeed, many teachers now recognise the significance of electronics as a cross-curricular activity, and realise the importance of electronics to all pupils.

The emphasis in the past has been to teach pupils the characteristics of components, whilst now we are much more concerned with a systems approach, that is, the way in which a group of components will act together to perform a useful task. This approach is further explored below.

The possibilities for using electronics within problem solving, the experiential learning approach advocated by TVEI, is only limited by the imagination. The Kent Electronics Kit will help teachers who are embarking on this relatively new approach in teaching methodology.

The work in the Pupil's Book has been divided into four sections:
1 Introductory Investigations
2 Transistor Investigations
3 Digital Electronics
4 Nuffield Extension Investigations

In the introductory investigations, pupils are introduced to the 'building bricks' of electronics through a systems approach. This work is ideal as an extension to the basic work on electricity, or as a stand-alone unit. The transistor investigations quickly progress to practical circuits using the transistor as a switch, and its use with a range of input sensors, which will have been studied in the introductory investigations.

In the investigations based on digital electronics, truth tables are investigated, leading on to more significant work which shows how logic gates can be used in decision making. The final section 'Nuffield Extension Investigations', is designed to allow pupils to experience more complex problems. The inclusion of a binary counter and a 7-segment display allows Nuffield GCSE Physics work to be undertaken.

Throughout the Pupil's Book there are many examples of practical applications for the circuits investigated, and the pupils' questions are open-ended. Suggestions for homework and further work are included in this Teacher's Guide.

Sensors	Switching devices	Output devices	Other components
LDR	Transistor	LED	Diode
Thermistor	Logic gates	Relay	Potentiometer
Reed switch	Op amp*	Buzzer	Capacitor
		Motor	Photodiode (Northern)
Nuffield Extension			
4-bit binary coding and decoding 7-segment display Binary counting and application The integrated circuit memory			

* Welsh board only at higher level.

The above table gives a general breakdown of electronics within GCSE physics. It is the interaction of these components that is of significance, not their characteristics. Whilst ten years ago it was normal to plot characteristic parameters, now it is considered that electronics should be taught as a *systems* approach. This means that a component should be studied as part of a larger circuit. It should be remembered that a system could be composed of only two components; a common example is shown below.

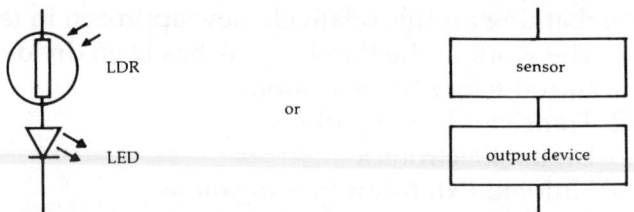

A simple light-dependent system

In this method of teaching, pupils will become aware of the practical implications and how an electronics circuit can be compartmentalised. For example, shown below is a burglar alarm treated as a system.

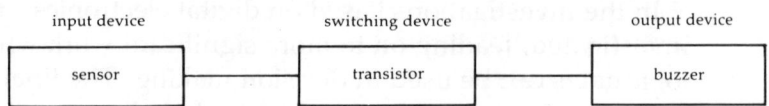

Many teachers initially find this black box 'jigsaw' approach to be unacceptable. However, experience has shown that as electronic circuits become more complex, this is the only sensible method of approach. How a circuit can be used and its practical applications are much more important concepts than how a particular component works.

The objectives of this course are for pupils to:
1   gain an understanding of basic electronics and its applications;
2   recognise components used in electronic circuits;
3   recognise the operation of a number of electronic systems;
4   adapt electronic circuits to give variation in operation;
5   experience problem solving situations;
6   understand basic applications of electronic systems;
7   become aware of the impact of electronics on society;
8   develop an interest and enjoyment in electronics.

COMPONENT RECOGNITION

Beginners in electronics often experience difficulty in identifying
components. Individual components are usually referred to as *discrete*
components; this section will help in their identification.

Resistors

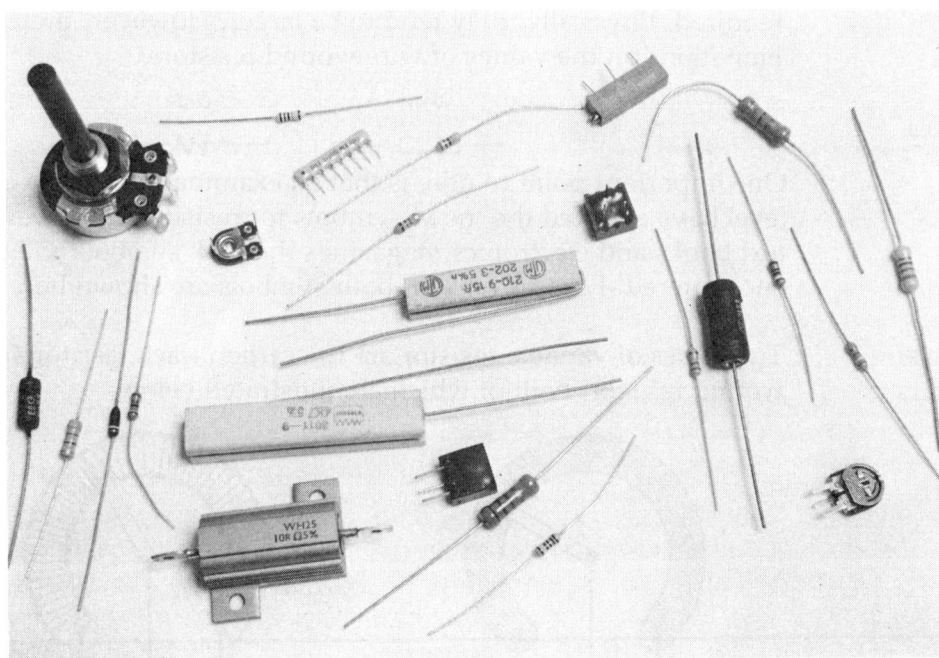

Uses

Resistors are used to:
1 limit current;
2 produce a change in voltage;
3 set a voltage level.

Units

The unit of resistance is the ohm:
ohm (Ω) 10^0 ohm
kilohm (kΩ) 10^3 ohms
megohm (MΩ) 10^6 ohms

There are several ways of expressing resistor values, for example a 4700 Ω resistor could be written 4.7 kΩ or just 4.7K. The BS 1852 resistance code employs the letters R, K and M as follows:

1 Ω is written 1R0	1 kΩ is written 1K0
4.7 Ω is written 4R7	2.7 kΩ is written 2K7
100 Ω is written 100R	10 kΩ is written 10K
	1 MΩ is written 1M0

Although all values of resistance from one to ten megohms are theoretically available, in practice the electronics industry has settled on a set of *preferred values*. These are called the E12 and E24 series.

Fixed
Resistors

Fixed resistors are constructed in several different ways, using carbon, carbon film, metal oxide or wire.

In general, carbon resistors are very cheap but are relatively unstable, as temperature fluctuations cause the value of the resistor to change. Carbon film and metal oxide resistors are much more stable and are less susceptible to a drift in value due to temperature change. Wire wound resistors have specialist uses, especially where a larger current is required. Physically, they tend to be larger. However, there are limitations on the values of wire wound resistors.

Symbols

One important point to note is that all examination boards at GCSE level have adopted the 'new' symbols for resistance. However, in many text books and electronics magazines the 'old' symbols will often be encountered. For this reason, both symbols are shown here.

Variable
Resistors

Two types of variable resistor are the carbon track resistor and wire wound resistor, both of which are illustrated below.

Symbols

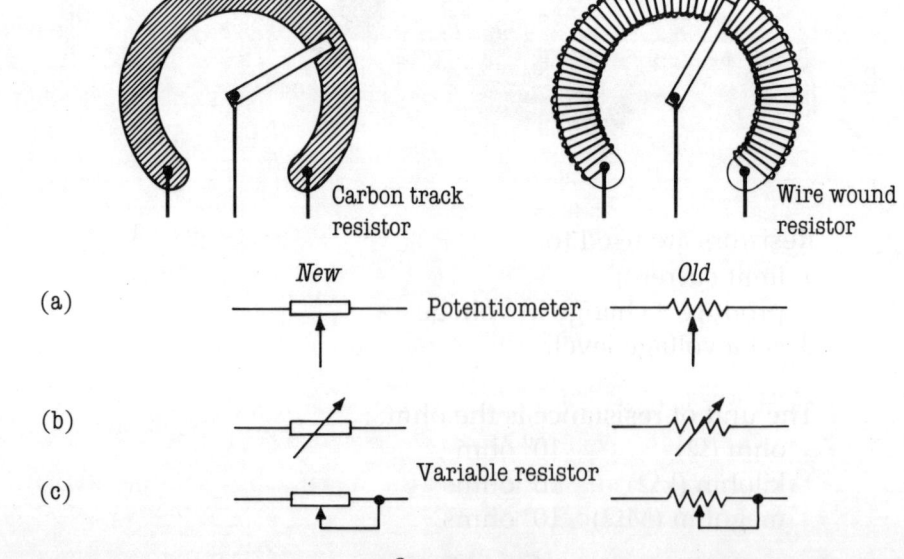

8

Power Rating

All resistors, whether fixed or variable, have a power rating in watts. If this power rating is exceeded, the resistor will overheat. It is very important to select a resistor that has an adequate power rating. The power dissipated by a resistor (in watts) can be found by multiplying the voltage across it by the current flowing through it. Physically large resistors generally have a higher power rating than smaller ones.

The Colour Code

The identification of a resistor value is achieved by means of the resistor colour code. This is a system of coloured bands, as shown on the adjacent diagram, which is used to identify the numerical values and the tolerance of the resistor. The table below lists the values indicated by the different coloured bands.

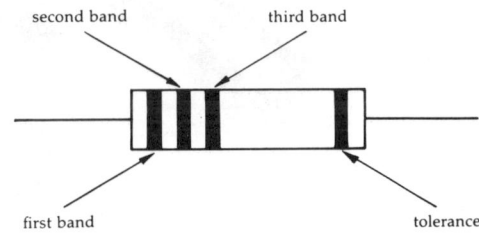

Colour	Band 1 1st figure	Band 2 2nd figure	Band 3 Multiplier	Band 4 Tolerance
Black	0	0	×1	
Brown	1	1	×10	1%
Red	2	2	×100	2%
Orange	3	3	×1000	
Yellow	4	4	×10 000	
Green	5	5	×100 000	0.5%
Blue	6	6	×1 000 000	0.25%
Violet	7	7	×10 000 000	0.1%
Grey	8	8		
White	9	9		
Gold			×0.1	5%
Silver			×0.01	10%
None				20%

The following example shows how the resistor colour code works.

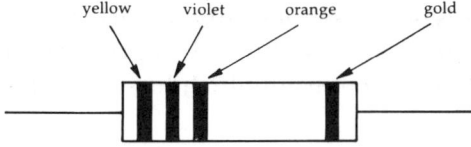

Reference to the table will give us the following information:

Yellow	Violet	Orange	Gold
4	7	000	5%

We can therefore deduce that the resistor has a value of 47 000 ohms or 47K, and a tolerance of ±5%.

Capacitors   A capacitor is a component that can store charge.

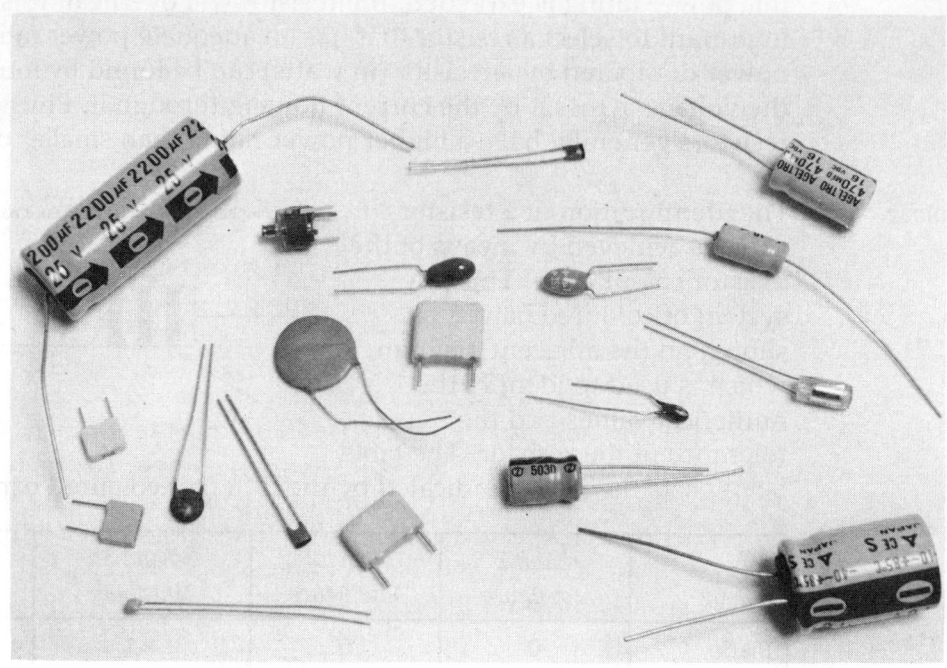

Symbols

Fixed capacitor   Variable capacitor

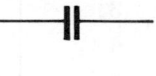

polarised

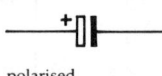

preset

Units   The unit of capacitance is the farad. The farad is a very large unit and therefore capacitors normally have values which are much smaller:

microfarad (μF) 10^{-6} farad
nanofarad (nF) 10^{-9} farad
picofarad (pF) 10^{-12} farad

Working   All capacitors have a maximum working voltage. In the case of
Voltages   electrolytic capacitors, this is marked on the capacitor itself. This voltage is the maximum that may be applied to the capacitor leads. A capacitor should *never* be used which has a smaller working voltage than is needed.

Construction   The simplest type of fixed capacitor is two metal plates separated by air, as illustrated opposite.

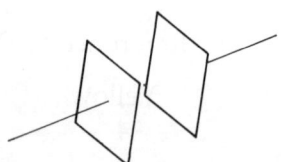

In this case, air acts as an insulating medium between the plates. This insulating medium is called the *dielectric*. Practical capacitors have a whole range of dielectrics, including polyester, polypropylene, silver mica and polycarbonate.

Electrolytic Capacitors

These are larger value capacitors mounted in an aluminium can, as shown below.

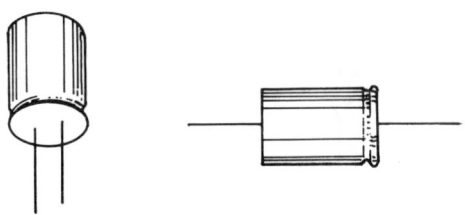

Electrolytic capacitors are polarised and must be connected the correct way around with respect to positive and negative. The dielectric in this type of capacitor is a layer of aluminium oxide. This method of construction allows large values to be 'housed' within a fairly small volume. One of the classic uses for an electrolytic capacitor is in a power supply for smoothing (*see* page 20).

Variable Capacitors

Variable capacitors are often used in radio circuits for tuning. One type of variable capacitor consists of two sets of plates separated by air. One set of plates can move over the other, effectively varying the surface area and hence the capacitance.

Diodes

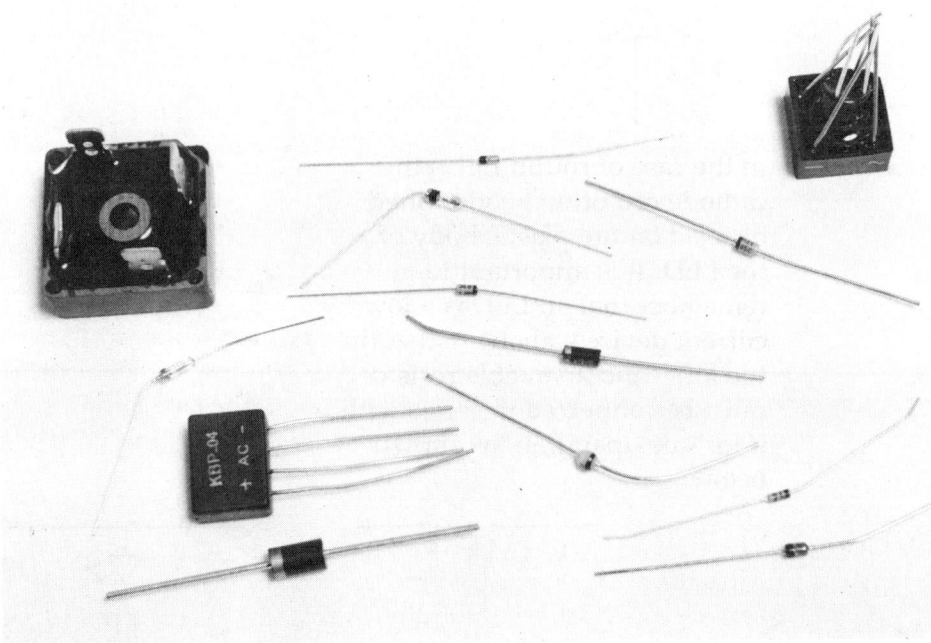

A diode is a semi-conductor device. It allows current to flow in one direction, but not in the other direction. When connected in the forward direction, the diode offers a low resistance and conducts the current. In the reverse direction, the diode has a very high resistance and will not conduct. Common examples of the types of diode available are the silicon diodes IN4001 and IN4148, and the germanium diode 0A91.

Symbols   The cathode is usually marked with a band.

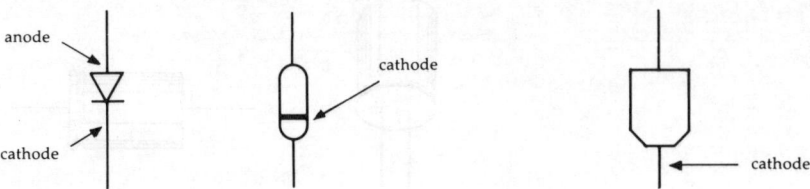

Types of Diode   Different types of diode are used to obtain various effects:

1 Point contact (germanium diodes) – often used in tuning circuits;
2 Junction (silicon diodes) – general use, e.g. rectification;
3 Zener – fixing a voltage.

Light Emitting Diode   A light emitting diode (LED) is a semi-conductor device made with a material called gallium arsenide. When connected in the forward direction the LED releases energy in the form of light. An LED will not function when connected in the reverse direction and may become damaged.

Symbol

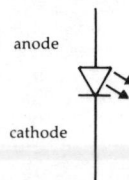

In the case of round LEDs the cathode can often be identified by a *flat* on the plastic body of the LED. It is important to remember that an LED is a low current device – about 10 mA (in the kit) – and a suitable resistor must be connected in series with it for safe operation, as shown below.

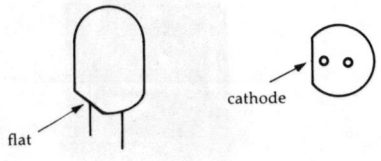

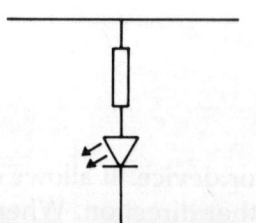
or
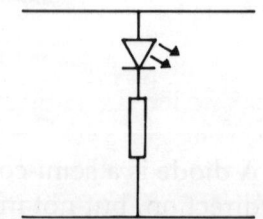

12

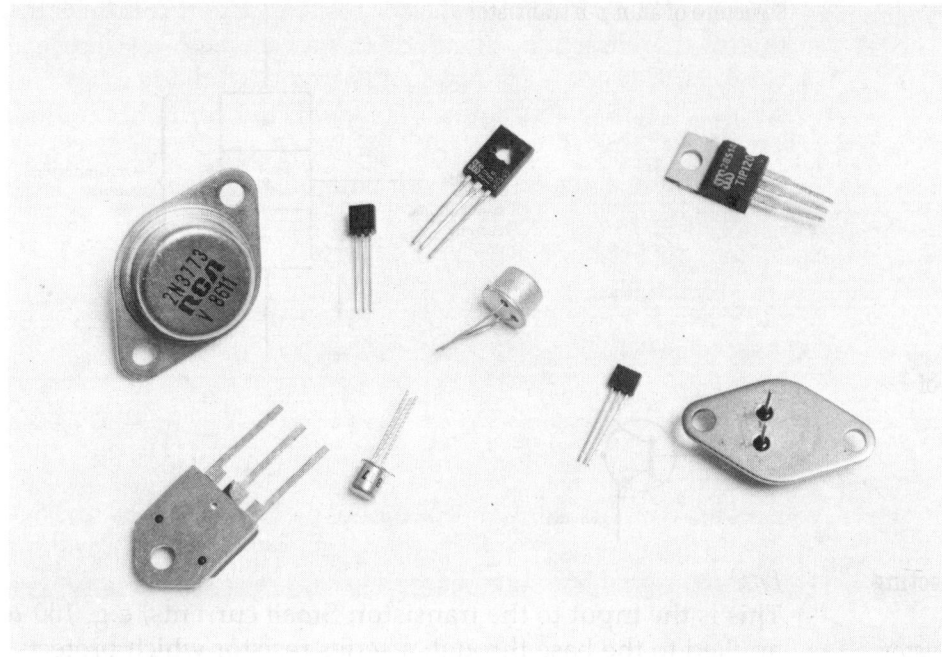

The transistor is a semi-conductor device. It is one of the most important 'building bricks' in electronics.

The transistor is a three-terminal device and great care is needed in the orientation of the transistor, as incorrect connection could result in damage.

There are two families of transistors, the *n-p-n* transistors and the *p-n-p* transistors, both of which are discussed under the headings below.

N-p-n
Transistors

Common n-p-n transistors	Approx. gain(h_{FE})	Maximum supply voltage	Collector current
BC 108	125+	30	100 mA
BC 109	240+	30	100 mA
BC 184L	200+	30	200 mA
ZTX 300	30	25	500 mA
2N 3055 power	20	70	15 A
BFY 51	40	30	1 A
TIP 31A	20	60	3 A

The *n-p-n* transistors are the more common type of transistor found in electronics today. All of the examples in the above table are silicon.

The letters *n* and *p* refer to the doping of the semi-conductor material. A semi-conductor material *n*-type has an excess of current carrying electrons, whilst a *p*-type has an excess of vacancies or holes which are regarded as having a positive charge.

13

Structure of an *n-p-n* transistor

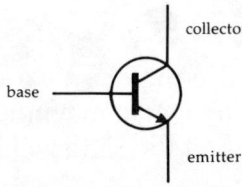

Symbol

base

collector

emitter

Connecting up a Transistor	*Base*

Connecting up a Transistor

Base
This is the input to the transistor. Small currents, e.g. 100 μA, are applied to the base through a series resistor which protects the input.

Collector
This is usually the output of the transistor and is connected to the + supply rail (V_{cc}) through a load resistor.

Emitter
This terminal is often connected directly to the zero volts rail, although in some applications the emitter can be the output of the transistor.

P-n-p Transistors

In general, the *p-n-p* transistor is not used as often as the *n-p-n* type. However, occasionally circuits arise that use both types simultaneously, as in the case of some audio amplifiers.

Structure of a *p-n-p* transistor

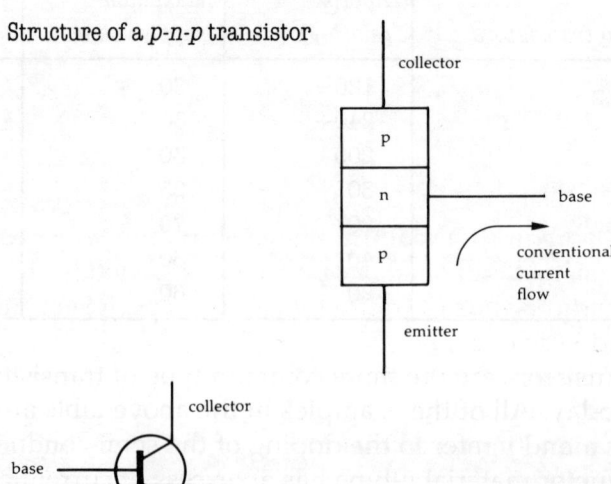

Symbol

base

collector

emitter

Carefully file off the top of a BC 108 transistor; alternatively the top may be removed with a junior hacksaw. Bend the legs at 90° to the case and mount on a glass slide, as shown below.

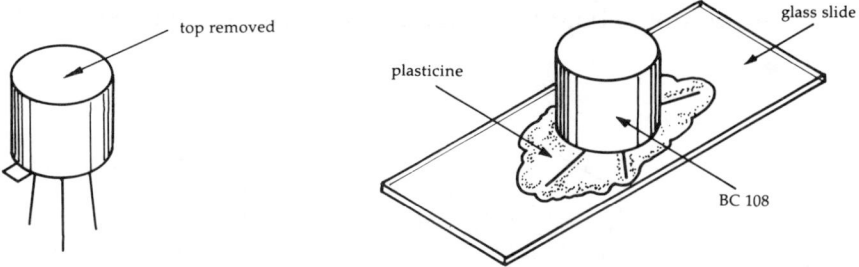

The silicon wafer and connecting wires can then be viewed under a microscope (remember to illuminate from above) at magnification × 5. This is an extremely worthwhile exercise, and the slide will keep for use in future years.

Integrated
Circuits

An integrated circuit (IC) is a tiny silicon chip onto which are etched transistors, resistors and small value capacitors. Modern ICs may contain many thousands of transistors and can perform complex tasks efficiently and very cheaply.

Integrated circuits have replaced large numbers of individual components and a typical IC may now require only a very small number of discrete components to make a complete circuit.

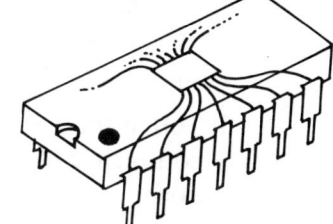

Most ICs that you will come across are mounted in a dual-in-line (dil) package. As the term implies, the pins are arranged in line, the distance between the pins being a standard 0.1 inch. This spacing allows ICs to be fitted to copper-clad stripboard (Veroboard) and also to breadboard or prototype board for circuit construction and experimentation.

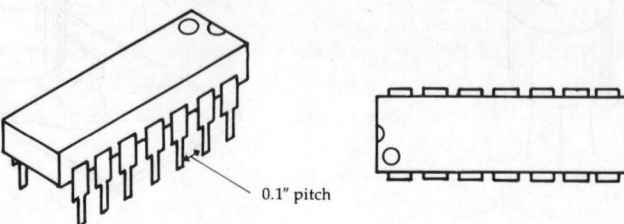

0.1" pitch

Dual-in-line packages may have 8, 14, 16 or more pins. The following are some typical examples:

741 operational amplifier (8 pins)
NE 555 timer (8 pins)
Logic gates (14 pins)
Microprocessor (64 pins)

Uses

There are many uses of integrated circuits, of which the following are some of the most familiar:

Digital ICs	*Analogue ICs*
Washing machines	Audio amplifiers
Computers	Televisions
Calculators	Tape recorders
Watches	
Compact disc players	

CMOS 74HC
Series

The Kent Electronics Kit uses the 'new' CMOS 74HC series of digital ICs. The following characteristics of these ICs should be noted, and taken into account when using them:

1 Voltage range 2–6 V dc
2 High-speed switching
3 Tiny current drain
4 Very high input resistance
5 Easily damaged by static electricity, unless fitted in a circuit board

Handling
CMOS
Integrated
Circuits

As the CMOS ICs are easly damaged by static electricity, they require special care when being handled. The following procedure should be adopted if it is necessary to fit or remove an IC.

1 Touch the *metal* casing of a power supply or similar appliance connected to the mains supply, to earth any static electricity in your body.

16

2 Remove the new IC from the anti-static sponge or transit package.

3 Insert the IC into its socket on the circuit board.

This procedure need not be followed when the kit is used normally, as the other components and wires on the circuit boards will 'earth' any static charge present.

LIST OF PUPIL INVESTIGATIONS

Introductory
Investigations

1 Regulator
2 Light Emitting Diode
3 Switch Position
4 'OR' and 'AND' Combinations of Switches
5 Reed Switch
6 Relay
7 Resistors
8 Potentiometer
9 Diode
10 Buzzer
11 Light Dependent Resistor
12 Thermistor
13 Capacitor

Transistor
Investigations

14 Transistor as a Switch
15 'Switch on' Voltage for a Transistor
16 Light Sensor
17 Temperature Sensor
18 Capacitor Delay
19 Driving a Relay
20 Combining Circuits 16(a) and 19

Investigations
Based on
Logic Circuits

21 Truth Table for a NAND Gate
22 Truth Table for a NOR Gate
23 Truth Table for a NOT Gate
24 Truth Table for a Buffer Gate
25 Truth Table for an AND Gate
26 Truth Table for an OR Gate
27 Using the Output as a Sink
28 Moisture Detector/Touch Switch
29 Light Sensor
30 Temperature Sensor
31 Simple Latch Switch
32 Reed Switch Alarm
33 Capacitor Delay
34 Driving a Transistor (and Relay)
35 Energy Saving Circuit Using a NAND Gate

36 Greenhouse Plant Waterer
37 House Alarm
38 Quiz Show Reaction Detector
39 Bistable Multivibrator
40 Monostable Multivibrator
41 Astable Multivibrator

Nuffield
Extension
Investigations

42 Binary to 7-Segment Display
43 Counting Single Pulses Using a Push Switch
44 Anti-Bounce Circuit
45 Counting Using a Light Beam
46 Astable Multivibrator and Counter
47 Astable Driving a Units and Tens Counter
48 Connecting a Logic Gate and LED

ORGANISING THE KENT ELECTRONICS KIT FOR CLASS USE

All class practical work requires care with the storage and distribution of equipment, particularly where technicians are scarce.

The boards making up the kit may be stored in cardboard boxes, although some schools have devised wooden containers with slots in which to place the boards. Assuming that ten (or more) sets have been purchased, the boards may be stored in the following ways:

(a) *PER SET*, where one box contains a complete set of boards;
(b) *PER BOARD*, where one box contains ten identical boards;
(c) a combination of both methods.

'Per Set'
Storage

This method reduces the problems associated with distribution, although care is needed in the layout of boards inside the box, to facilitate checking at the end of a lesson. If pupils normally work in the same positions in the lab, their group working positions could be numbered. If individual boxes were similarly numbered, and always used by the same group of pupils, there would be a greater incentive for pupils to take care of the apparatus, particularly as the teacher would be able to trace the pupils responsible for any loss or damage.

The main disadvantage of the 'per set' method is that pupils may be distracted or confused by the boards which are not being used in a particular lesson. This would be a greater problem for younger pupils, who may be following the 'Introductory Investigations'.

'Per Board'
Storage

This approach overcomes the problem, since the teacher is able to introduce pupils to a new component(s) when appropriate. However, distribution at the beginning of a lesson, and collection at the end, would need to be well organised and controlled.

Combination
Method

Many schools may opt for a compromise, where most boards are stored 'per set', but items such as the relay, transistor and logic gates are stored 'per board'.

The connecting links are stored on a purpose-made storage board and, providing that the storage boards are fully occupied with links, checking should be quick and easy.

Using the
Kent
Electronics
Kit

The printed circuit board modules are connected together by means of the special linking plugs supplied. Pupils should be encouraged to count out the number of linking plugs required for their circuit before connecting the boards together. Many circuits can be constructed without using any flying leads (e.g. flexible connecting leads). Where a flying lead is required, the best type to use consists of a short flexible wire fitted with a miniature (insulated) crocodile clip at each end.

The kit has been designed to achieve maximum flexibility, and individual boards can be made into almost any circuit arrangement which may be required, particularly if flying leads are used freely.

Power Supply
Requirements

The regulator provided with the Kit will operate on a supply of between 8 and 20 V dc or 8 and 16 V ac. Never exceed 20 V dc or 16 V ac. A 12 V dc supply is ideal, and the following are suggested sources of power:
> Low voltage lab pack or similar low
>   voltage supply;
> Low voltage distribution system
>   found in many labs;
> 12 V car battery (with a 1 A fuse in
>   series between the battery and
>   regulator);
> 9 V PP9 battery;
> 9 V PP3 battery (for short term use);
> 9 V battery eliminator mains unit.

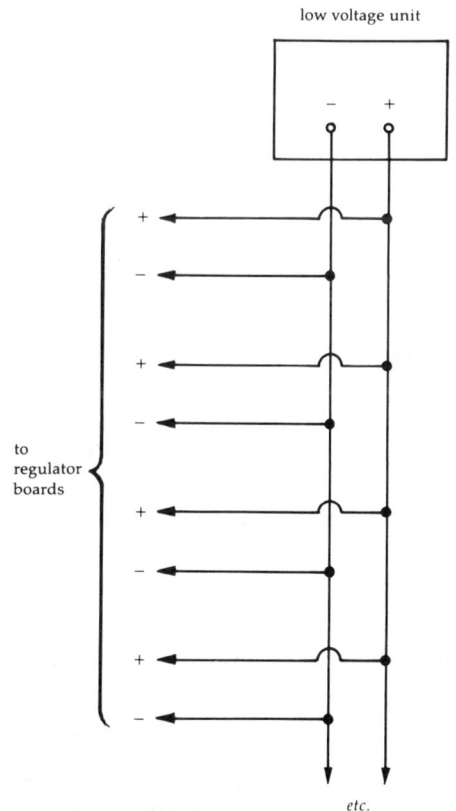

Although the regulator will work on an ac supply, where a choice exists, dc is recommended.

The regulator requires a maximum current of about 100 mA. Since most low voltage units used in labs supply 1 A or more, several regulators may be connected in parallel to the same supply. In fact many low voltage units are quite capable of driving up to ten regulators connected in parallel.

POWER SUPPLIES

**Simple Mains
Power Supply**

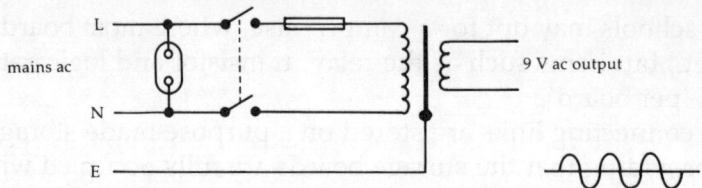

The above circuit shows a mains transformer with the usual mains safety devices in the primary circuit. The secondary coil provides (in this case) a 9 V ac nominal output. In practice, the output voltage may be higher if there is little or no output current, and will be lower if the output current rises above the maximum permitted. These changes in the output voltage may cause serious problems in electronic circuits, particularly if the load current used changes significantly.

**Root-Mean-
Square
Values**

Assuming that the transformer supplies 9 V, this value is the *root-mean-square* (rms) value. Thus it could be expressed as 9 V ac rms. The rms value of an alternating current (or voltage) is the steady direct current (or voltage) which would provide the same heating effect. For an ac sine wave (e.g. mains derived ac):

rms value $\quad$ = peak value$/\sqrt{2}$
$\qquad\qquad$ = peak value $\times$ 0.7 (approx.)
thus peak value = rms value $\times$ 1.4.

A 240 V dc supply will produce the same heat energy per second as a 240 V rms ac supply connected to an identical heating element. The peak value of a 240 V rms ac supply is actually 340 V.

Returning to our 9 V rms transformer, the 9 V rms ac output actually becomes a 9 $\times$ 1.4 = 12.6 V *peak*.

**Rectifying
The AC
Output**

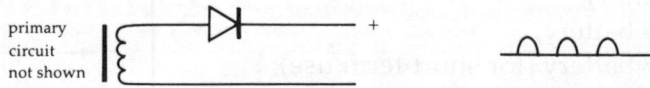

The above circuit shows how the ac supply may be changed into dc by means of a single diode. The output is a *half wave* dc supply, and is unsuitable for most electronic circuits, which require a smooth (straight-line) dc.

**Smoothing
The Output**

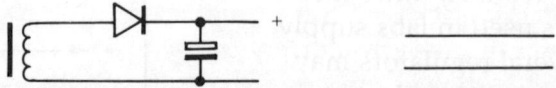

The capacitor shown above would normally have a large value and be capable of storing energy as the voltage rises to its peak, and then releasing some of the energy as the voltage falls towards zero.

Under 'no load' conditions, the dc output will rise to the peak ac value, and will remain at that level. When a 'load' is connected, the output voltage will 'ripple' as shown below.

Some electronic circuits will work satisfactorily on this type of supply, particularly if a very large value capacitor is used to reduce the ripple. However, large value capacitors are both expensive and bulky. A less expensive alternative is to provide a *full wave* dc supply.

Full Wave DC
Supply

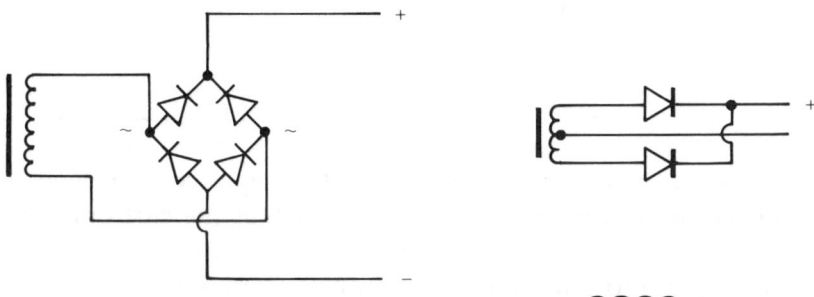

The circuits above show a full wave dc supply. The left-hand circuit employs four diodes, or a bridge rectifier (which consists of four diodes in one package). Both the positive and negative cycles of the ac output are now conducted via one diode or another, producing a full wave output.

The right-hand circuit employs two diodes and a transformer with a *centre tapping*. In other words, the secondary coil has a third connection at its centre point.

In both cases, a smoothing capacitor would now have an easier task, because the crests of the wave are closer together. The dc supply obtained is suitable for many electronic circuits, providing that they are fairly tolerant to voltage fluctuations.

WHY USE A REGULATOR?

Unregulated power supplies have two main disadvantages:

1 the output voltage varies with the current used;
2 incorrect wiring can cause the current to rise to a sufficiently high level to damage the power unit, or the circuit, or both.

A regulated power supply will maintain the required voltage regardless of the current taken, unless this exceeds the maximum permitted, at which point the output voltage should fall towards zero.

In the case of the Kent Electronics Kit regulator board, this is indicated by the LED, which stops glowing.

Integrated Circuit Regulators

A wide range of IC voltage regulators is now available. For convenience we will consider fixed voltage 3-pin devices. Even these appear in a variety of shapes and sizes, and fall into two groups: *positive* regulators and *negative* regulators.

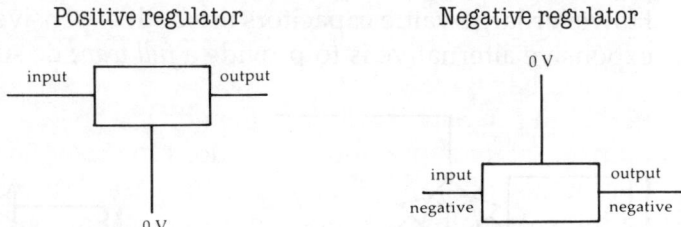

The positive IC regulator allows the negative lead to be 'grounded', i.e. connected to an earthing point such as the earth pin of a mains plug. Thus the negative supply becomes 'zero volts'. It is often convenient to call the negative supply 'zero volts', even if it is not connected to earth.

The negative IC regulator allows the positive lead to be 'grounded'. Thus the positive supply can become 'zero volts'. This type of IC may not be particularly useful in single rail supplies, but is often required (together with a positive regulator) in a dual supply, as shown below. This type of supply shown in the diagram is often required in operational amplifier (op amp) circuits.

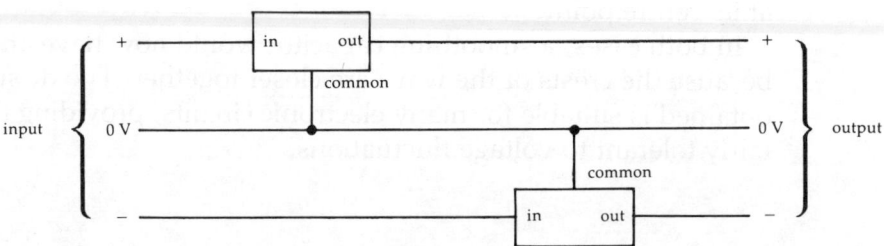

Types of Voltage Regulator IC

The 100 mA regulator IC used in the Kit is a type 78L05. The second digit, '8', indicates that the IC is a positive type. The equivalent negative IC is type 79L05. The last two digits, '05',indicate the voltage output (plus or minus 4%). The letter 'L' indicates a maximum current of 100 mA. Thus the IC provides a 5 V supply at any current up to a maximum of 100 mA.

Input Voltage

In each case, the input voltage to the IC must be 2–3 V more than the voltage the IC is designed to supply. The maximum input voltage is between about 25 V and 40 V depending on the type.

Protection

All the regulators described above incorporate 'short circuit protection' and 'thermal shutdown'. They cannot therefore be harmed by a short circuit across the output, and if they overheat for any reason, the output voltage is reduced until the IC becomes cool enough to operate properly again.

Other Varieties

Various other types of IC regulator are available, some with more than three leads, offering variable voltage and current limiting. A good quality catalogue will indicate the wide variety available.

The Kent Electronics Kit Regulator

Modern IC regulators are so efficient that even a substantial ripple at the input can be virtually eliminated. The smoothing capacitor shown on page 20 may therefore be reduced in size. The consequent saving in cost may often be greater than the cost of the regulator IC. An additional small-value capacitor can be connected between the input and 'ground' to reduce any voltage spikes which may appear on the supply. A similar small value capacitor is required between the output and 'ground'. The full circuit, as fitted on the regulator board, is shown below.

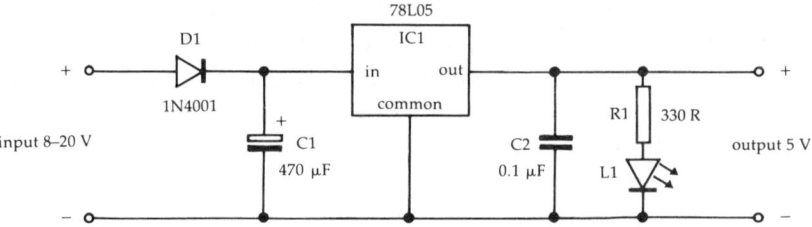

Note: An additional small capacitor on the input side is not required in this circuit.

Investigation 1
Regulator

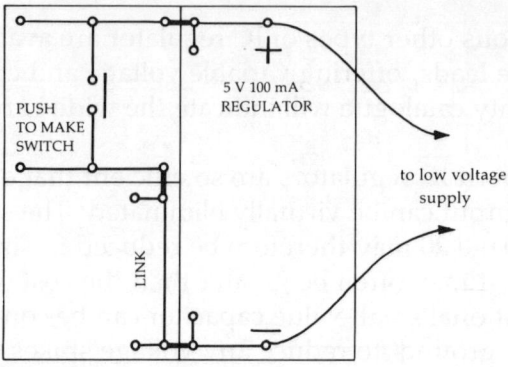

PUSH
TO MAKE
SWITCH

5 V 100 mA
REGULATOR

+

to low voltage
supply

LINK

−

Circuit 1

When connected to a low voltage supply, the regulator provides a 5 V dc supply. The positive output connection is denoted '+5 V' and the negative connection should be called '0 V'. This convention is helpful in later experiments, where voltage measurements are taken at various points throughout a circuit. If the voltmeter negative lead is connected to the negative or 0 V line, the positive voltmeter lead may be used as a 'probe' to measure voltages at various points, for example at a transistor base connection. The reading on the voltmeter will then be the voltage at that point in the circuit, relative to the 0 V line.

The first investigation shows pupils what happens if a short circuit occurs across the output of the regulator. When the push switch is pressed, the regulator LED stops glowing. This indicates to pupils that a serious fault exists, and the power unit should be switched off as soon as possible.

The regulator included in the Kent Electronics Kit will withstand a short circuit for an unlimited time; however, it is still essential to stress to pupils that short circuits should be avoided.

The regulator will prevent damage caused by virtually all the possible wiring mistakes which pupils may make – even connecting a board (such as the NAND board) with the wrong polarity (i.e. upside down).

It should be stressed, however, that pupils must *never* connect a flying lead to any metal point on the regulator board itself, as such action could lead to a short circuit across the power unit supplying the regulator, or could allow the 12 V (or more) input supply to reach the electronic circuit.

Investigation 2
Light Emitting Diode

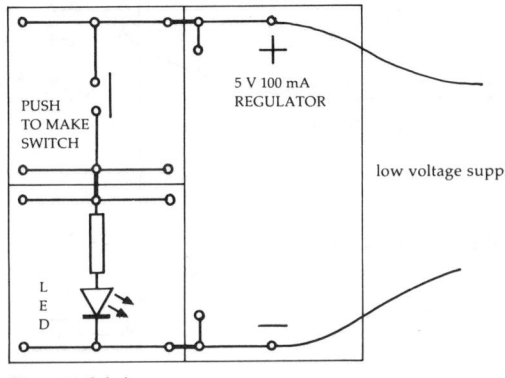

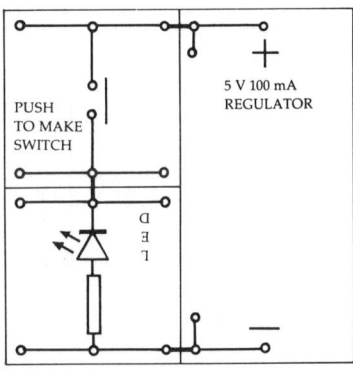

Circuit 2(a)　　　　　　　　　　　　Circuit 2(b)

The investigations demonstrate that an LED must be connected in the 'forward direction'. Reversing the LED will prevent it lighting, and may even cause it some damage, if the supply exceeds about 5 V.

An LED offers very little resistance to current in the forward direction, and a series resistor is essential to limit this current to a safe level. The value of this resistor is calculated using Ohm's Law, noting that the LED used in the Kit requires about 10 mA, and maintains when conducting a potential difference (pd) from anode to cathode of about 2 V. Thus on a 5 V supply a potential difference of $5 - 2 = 3$ V will exist across the series resistor.

$$\text{Using Ohm's Law, } R = V/I$$
$$\text{thus } R = 3/0.01 \text{ (where 0.01 A = 10 mA)}$$
$$= 300 \ \Omega.$$

It may be noticed that a value of 220 Ω has been chosen for the Kit. With a pd across the resistor of 3 V, the current provided is as follows:

$$I = V/R$$
$$\text{thus } I = 3/220$$
$$= 0.014 \text{ A or 14 mA.}$$

This larger current is used to ensure good results even in a bright classroom.

LEDs may be operated on an ac supply. The usual method is to connect a diode in reverse parallel as shown in the diagram on the right. The value of the series resistor may be reduced to allow for the fact that current flows through the LED in pulses for only half the time.

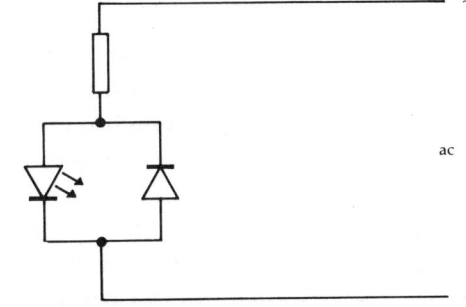

Further Work	Pupils could be asked to connect a red LED and a green LED in 'inverse parallel' as shown in the diagrams. Using a pair of flying leads (e.g. flexible leads fitted with crocodile clips), connect the LEDs to the regulator with the polarity first one way, then reversed, as shown.

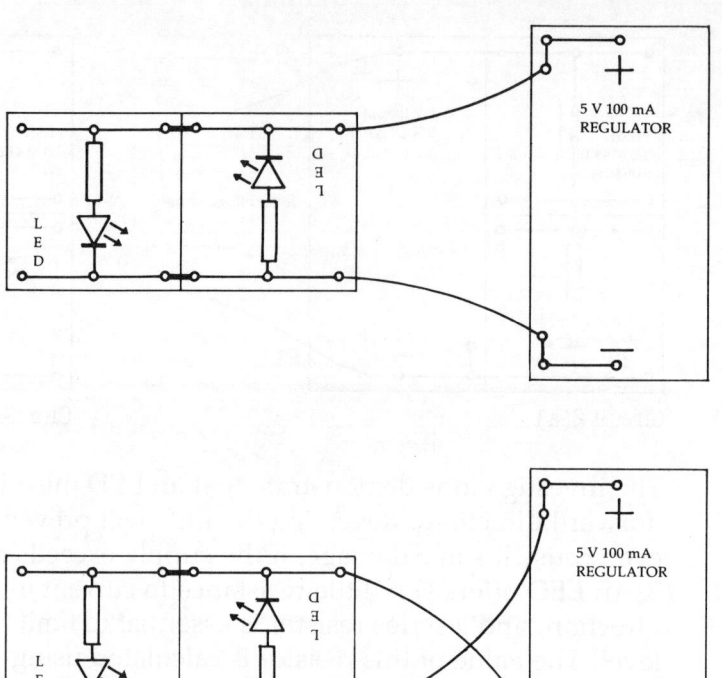

This circuit illustrates how a pair of coloured lights (e.g. model railway signals) can be changed from red to green simply by reversing the supply. The circuit is useful since only two wires are needed between the power supply and the two LEDs. It is possible to purchase a red/green LED which appears to be a single LED but is like the circuit shown above and glows red with the current flowing in one direction, and green when the current is reversed.

Homework Suggestion	Pupils could be asked to list all the LEDs (including arrays in digital number displays) which they can find at home. *Note*: Seven-segment LEDs, which are normally red, green, or yellow illuminated segments making up a number, should not be confused with liquid crystal displays, which comprise black segments on a silver, or sometimes coloured, background.
Applications	Light emitting diodes are virtually everlasting and use very little current. They generate almost no heat and are used extensively as indicators, for example in video recorders. Their low power

consumption makes them ideal for use in battery-powered equipment. They are available in red, amber, yellow and green, and in a variety of shapes and sizes. They are also used in 'straight-line arrays' to produce bar graphs and audio-level indicators. Seven-segment arrays are commonly used to form illuminated numbers.

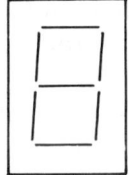

Investigation 3
Switch Position

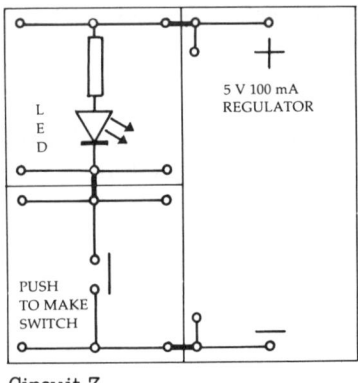
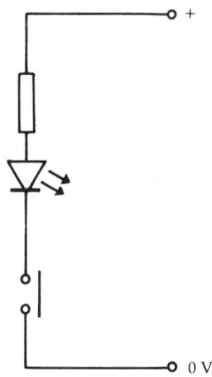

Circuit 3

The circuit illustrates that a switch must be wired in series with the device it is controlling. It is important to note that the switch may be placed either on the positive *or* the negative side of the device being controlled. Many pupils fail to understand this. The circuit deliberately places the switch on the negative side of the LED, unlike Circuit 2(a) where the switch is connected to the side many people consider to be the 'normal' position. The circuit is important because the transistor circuits described later place the transistor (which acts as a switch) on the *negative* side of the device being controlled.

It is essential to note that the above paragraph applies only to dc circuits. When switching, for example, mains ac, *both* the *live* and *neutral* lines should ideally be controlled by a switch. However, if a single-pole switch has to be used, it *must* be wired on the *live* side of the circuit. Many people confuse the *positive* and *negative* sides of dc with the *live* and *neutral* lines of mains ac. It should be remembered that a live ac line changes from positive to negative and back again many times a second. The neutral line remains at about zero volts.

The switch included in the kit is a push to make, single-pole type, similar to a bell push. This is sometimes described as a momentary switch, since the switch is only closed (on) whilst pressure is applied. Another type is the 'latching' switch. In this case the switch latches ON and remains ON after being pressed, and latches OFF when pressed again.

Background
Information

Another type of momentary switch is the 'push to break' type. This is normally closed (on), and when pressed the switch opens (switches off).

Many other types are available. Very common is the 'toggle switch'. These can be 'closed' or 'opened' and will then (normally) remain in that state (e.g. a domestic light switch). They may have more than one pole in order to control more than one circuit at a time. For example, toggle switches in a mains socket should have two poles in order to switch off both the live and neutral lines.

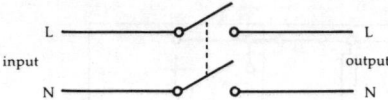

Toggle switches may have more than one 'way', and varieties of combinations of 'ways' and 'poles' are available.

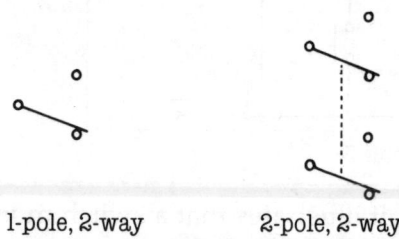

1-pole, 2-way   2-pole, 2-way

Other varieties of switches which can offer a similar function include rocker switches and slide switches. Rotary switches are available which offer a large number of 'poles' and/or 'ways'.

Applications
and Homework

Pupils could be asked to locate and list the types of switches which they can find at home, particularly the less obvious ones, such as the keys of a computer keyboard, or the handset switch in a telephone.

Investigation 4
'OR' and 'AND' Combinations of Switches

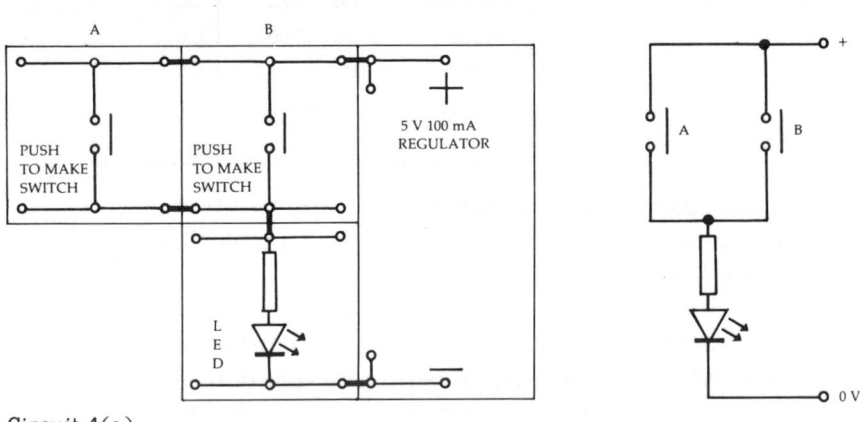

Circuit 4(a)

The 'OR' combination of switches, Circuit 4(a), shows how two switches may be wired in parallel to control the same device, where one switch, *or* the other, *or* both, may be pressed to switch on the device. The truth table results should be as follows:

Switch A	Switch B	LED
Released	Released	Off
Pressed	Released	On
Released	Pressed	On
Pressed	Pressed	On

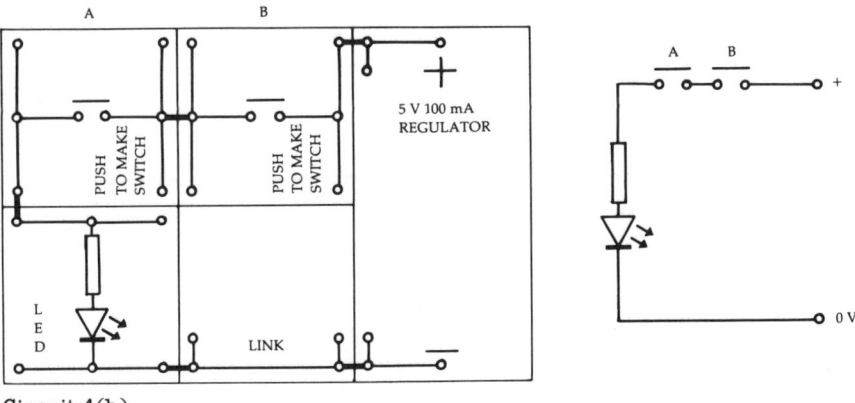

Circuit 4(b)

The 'AND' combination of switches, Circuit 4(b), shows how the switches may be wired in series to control a device. This time, one switch *and* the other must be pressed at the same time in order to switch on the device.

The truth table results should be as follows:

Switch A	Switch B	LED
Released	Released	Off
Pressed	Released	Off
Released	Pressed	Off
Pressed	Pressed	On

These circuits could offer pupils an introduction to logic; the theory surrounding this topic (including truth tables) will be dealt with later (page 79).

Applications

A possible application of an OR switch arrangement includes a door bell, where a push switch is provided at both the front and back doors. Pushing either switch (or both) will ring the bell.

The AND switch arrangement may be found in the safety device on a tape recorder, where *both* the 'play' *and* 'record' buttons must be pressed at the same time, before the machine will record.

Further Work

Pupils could borrow additional switches from other kits, and connect three or more switches in parallel or series. A house alarm system may have three or more 'under-carpet pressure mats' (which operate like push to make switches), wired in parallel. The same alarm system may also have three or more switches in series, to protect doors and windows. Such switches are the 'normally on' types. In other words they are on, unless an intruder enters, at which time one will switch off, breaking the series circuit. This system is discussed further in Investigations 5 and 6.

Investigation 5
Reed Switch

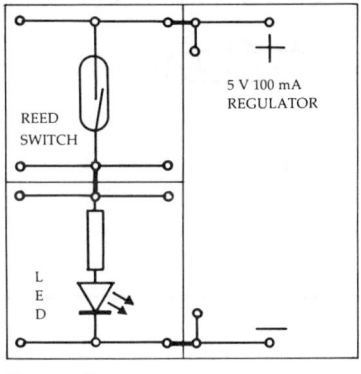

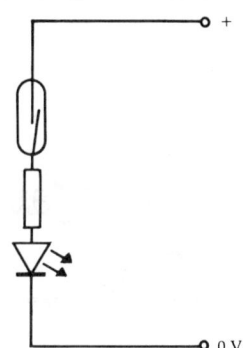

Circuit 5

The reed switch supplied with the kit is a 'normally open' (i.e. normally off) switch. When a magnet is placed near the reed switch, the contacts become magnetised as shown below, and the opposite magnetic poles attract, causing the switch to 'close' (switch on). In order to obtain the best results, the axis of the magnet should be placed in line with the reed switch.

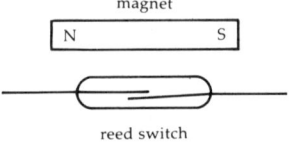

Further Work

Pupils could be asked to fix a small magnet to a pendulum bob (or use a suspended magnet as a pendulum). This should then be allowed to swing just above the reed switch, taking care *not* to let the magnet strike the reed switch, which could be protected with thin card. Each time the magnet passes the reed switch, the LED will flash on and off. This effect could be used to help count the pendulum swings, or used with an automatic counter or timer.

Applications

Reed switches are frequently used in burglar alarms to protect doors and windows. The reed switch is placed on (or often in) the frame of the door or window. A magnet is then fixed to the door or window in such a way that, when shut, the magnet is next to the reed switch. The reed switch is therefore 'closed' (on). When the door or window is opened, the magnet moves away from the reed switch, and its contacts open. A suitable electronic circuit, which sounds an alarm, is described later (Investigation 37).

Reed switches may be placed inside a coil of wire. When current flows through the coil, the magnetic field produced causes the reed switch

31

contacts to close. Such a device is known as a *reed switch relay*. The current required to operate such a relay is normally less than that required for a standard relay. However, the maximum current that can be controlled by the small reed-switch contacts is lower than that (normally) controllable by a standard relay.

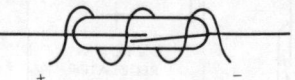

Investigation 6
Relay

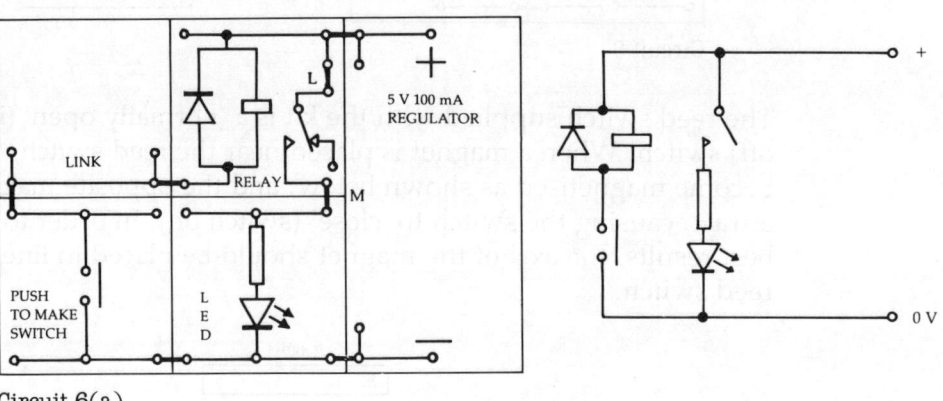

Circuit 6(a)

Circuit 6(a) demonstrates that a relay may be used to switch on a device, such as an LED, or a buzzer. The circuit may not appear particularly useful to pupils, though it should be noted that the relay is able to switch on a much larger current than the current which flows through the push switch.

Background
Information

The relay consists of a coil wound around a soft iron core. When an electric current flows through the coil, the soft iron core becomes temporarily magnetised, and attracts an iron armature. The armature rocks on its pivot and operates a switch contact. If the current through

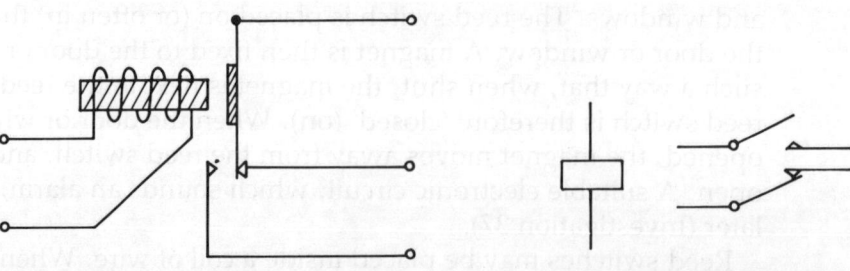

Simple diagram of relay   Relay with changeover contact

the coil is switched off, the armature is no longer attracted and returns to its normal position.

In practice the armature may control more than one set of contacts, and such contacts may be 'normally open' (i.e. normally off), 'normally closed' (i.e. normally on), or 'changeover' types, as shown below.

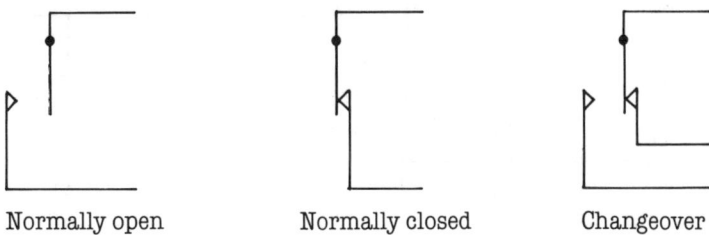

Normally open   Normally closed   Changeover

The relay used in the Kent Electronics Kit has a single 'changeover' contact, which can be connected as a 'normally open' or 'normally closed' contact.

If link L in Circuit 6(a) is removed, the centre contact of the relay may be connected to any external voltage (providing that it does not exceed the rated maximum of the relay contacts). Thus the normal 5 V supply may be used to control a different voltage, e.g. 12 V. In this case the 0 V line is common to both supplies; in other words the negative side of the 12 V supply is connected to the negative side of the 5 V supply.

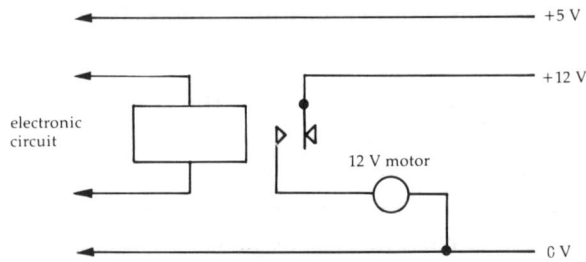

If the LED in Circuit 6(a) is disconnected, the relay contacts are completely isolated from the other components and the supply. Providing total isolation is an important application of a relay. For example, a mains lamp (as shown below) or motor could be switched via a relay, with no danger of the mains supply reaching the associated electronic circuit (providing that the relay is correctly wired). *Note*: Pupils should *never* experiment with the mains electricity supply.

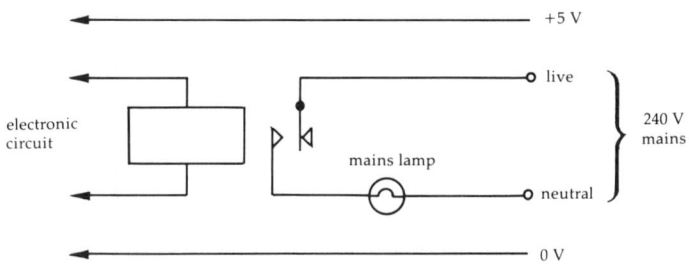

Circuit 6(b) below shows how the 'changeover' contact may be used as a 'normally closed' (i.e. normally on) contact. The LED should light when power is applied to the circuit, but when the push switch is pressed the LED switches off.

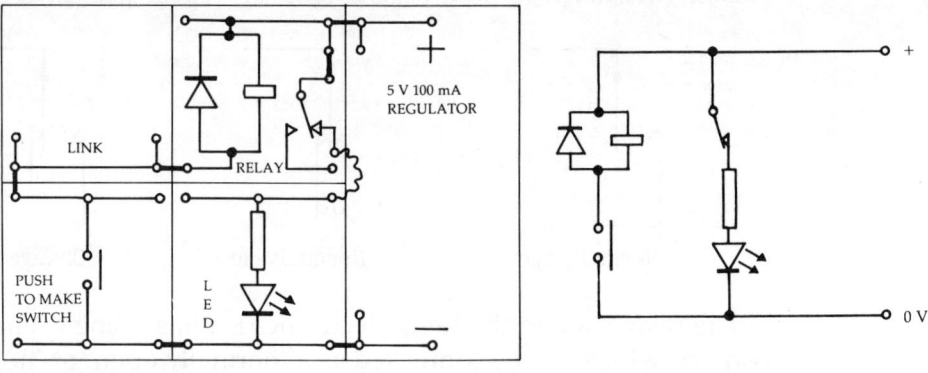

Circuit 6(b)

When selecting a relay, the following points should be taken into account:

1 coil operating voltage;
2 coil resistance. Ensure that this is *not lower* than that specified in the circuit diagram (a higher resistance is perfectly acceptable);
3 number and type of switch contacts offered;
4 maximum voltage rating of the relay contacts;
5 maximum current rating of the relay contacts.

When a relay switches off, the collapsing magnetic field induces a high voltage across the relay coil. This is known as *back emf* and may be high enough to damage the transistor or other device driving the relay. Fortunately this back emf can be 'shorted out' by means of a diode connected with its *cathode* towards the *positive* side of the relay coil. Almost any cheap silicon diode may be used, such as the types 1N4148 or 1N4001. It is essential to connect the diode the correct way round, as shown below. A diode facing the wrong way may conduct a sufficient current to destroy the transistor or other device which is driving the relay. However, this should not be a problem with the Kit, since the diode is already wired into the relay board, and the regulator should also provide sufficient protection.

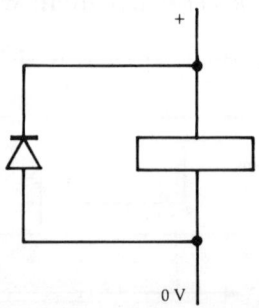

Further Work   A house alarm system which employs reed switches to protect doors and windows has already been mentioned (*see* page 31). If a reed switch is placed in series with a buzzer, the buzzer will sound when the magnet is next to the reed switch, i.e. when the door or window is closed. Pupils left to experiment with this circuit will, hopefully, realise that this is the wrong way round, i.e. the alarm sounds when the door is shut, and stops sounding when the door is opened.

If the reed switch is connected in series with the relay coil, and the buzzer connected to the 'normally open' contact, the alarm will still work the wrong way round. The only advantage gained by using the relay is that a very small current flowing via the reed switch and relay coil can turn on a much larger current to work a loud siren via the relay contacts.

However, some pupils may then be able to discover how the circuit can be modified so that the buzzer sounds when the magnet is moved away from the reed switch (i.e. the door is opened). The solution is to use the 'normally closed' contact as shown in the diagram below.

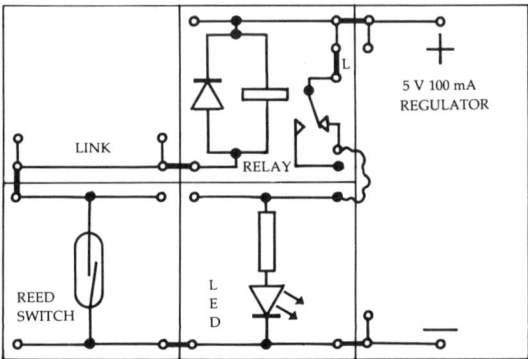

The alarm circuit is still less than perfect, since an intruder could quickly close the door or window and stop the buzzer sounding. A cunning alteration to the circuit, shown in the diagram below, makes the buzzer continue to sound, even when the magnet is replaced against the reed switch.

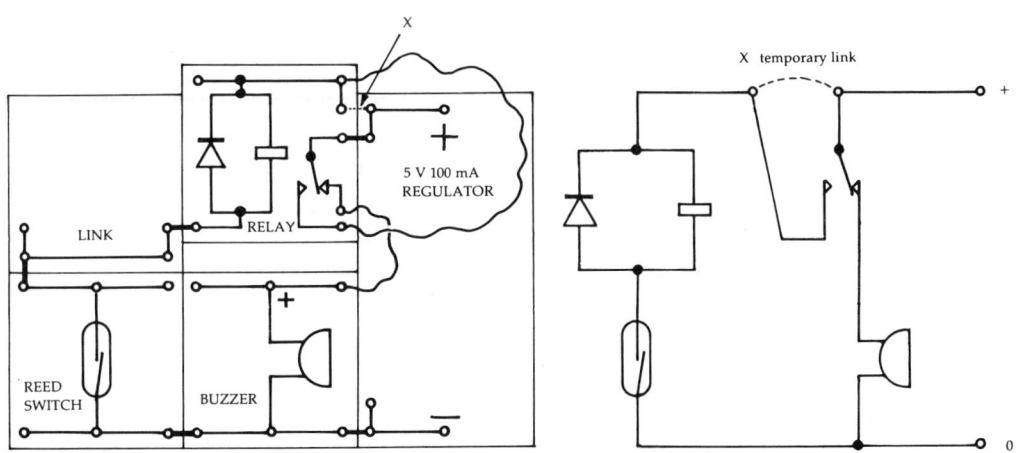

35

The link marked 'L' is removed, and the centre contact of the relay is connected directly to the positive supply, thus isolating the top connection of the relay coil. A second flying lead is then added, to connect the top of the relay coil to positive via its 'normally open' contact.

The magnet should be placed against the reed switch. When the power supply is switched on, the buzzer will sound. Insert link 'X'. The relay will switch on, and the buzzer will stop sounding. Now remove link 'X'. The relay will remain switched on, because current will flow to the relay coil via its switch contacts.

Move the magnet away from the reed switch (e.g. the intruder enters by opening a door). The relay should switch off, causing the buzzer to sound. Replace the magnet (e.g. the intruder closes the door). The buzzer should continue to sound as current can no longer reach the relay coil via its contacts.

Homework Suggestion	Pupils could be asked to draw the circuit of a complete alarm system based on the simple or the more complex arrangement above, to protect their house, or just one room (e.g. the pupil's bedroom).
Other Applications	Numbers of relays can be interconnected to perform 'logical' tasks, for example in telephone exchanges. However, such tasks are now performed so efficiently by logic integrated circuits that relays are being phased out in this area. However, relays are used extensively as *output* devices, where a fairly small voltage and current from, for example, a transistor can be used to switch on a much higher voltage and/or current.

Investigation 7
Resistors

Circuit 7 demonstrates the effect of an additional resistor in series with the LED. The effect is to reduce the flow of current, the value of which can be calculated using Ohm's Law (*see* the section on LEDs on page 25 for clarification if necessary). Series resistors are frequently used for this purpose; another example is the resistor in series with the base connection of a transistor to limit the flow of current into the base.

The voltage, or potential difference (pd), across a resistor depends upon the value of the resistor (measured in ohms) and the current flowing through it. In many circuits the current may vary, and this will

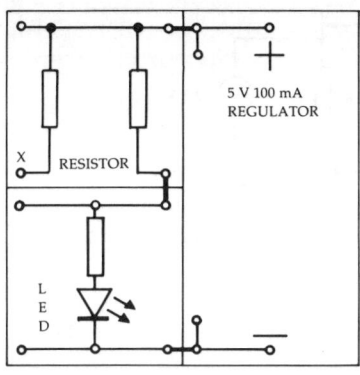 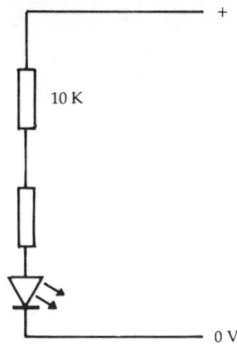

Circuit 7

cause the pd to change. For example, a 10 kΩ (10 000 ohms) resistor will produce no pd if the current flowing is zero:

$$V = I \times R$$
thus $V = 0 \times 10\ 000$
$$= 0$$

(where V is the pd in volts, I the current in amperes, and R the resistance in ohms).

The example shows that the pd is zero, regardless of the value of the resistor.

However, if a current of, say, 0.3 mA is flowing through the resistor, the potential difference (V) across the resistor will be given by:

$$V = I \times R$$
thus $V = 0.0003 \times 10\ 000$
$$= 3$$

Thus the pd across the resistor is now 3 V.

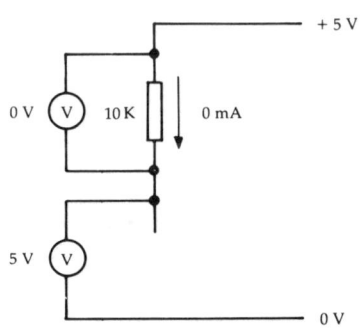 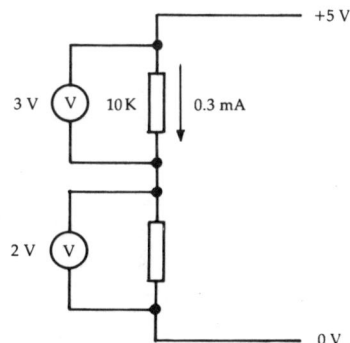

Potential Dividers

Electronic circuits often require different voltages at various points. This can be easily achieved using a pair of resistors as a *potential divider*.

If the two resistors are of equal value (say 10 kΩ), the voltage at the junction between them will be half the supply voltage (measured with respect to the 'zero volts' rail). Ohm's Law can be used to show this, and a high resistance voltmeter may be used to check the result.

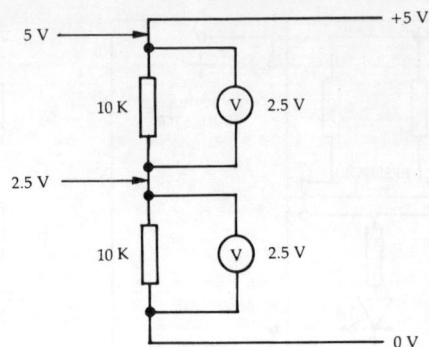

It is important to note that any current taken from the junction between the resistors will reduce the expected voltage. Thus a low resistance voltmeter may cause a noticeable error. In practice, the junction between the resistors is often connected to another part of a circuit. The maximum 'diverted' current required should be noted, and the resistor values chosen so that ten times this current flows through the potential divider. The total error resulting from the 'diverted' current will then be quite small.

In a practical circuit the values of the resistors should not be too low, or the flow of current through the potential divider (which is effectively wasted energy) may cause overheating or – in the case of battery-powered equipment – a shorter battery life. Clearly, the current diverted away from the junction should be as low as possible. CMOS integrated circuits (the type used in the Kent Electronics Kit) offer the considerable advantage of an almost infinite input resistance. Thus, when connected to a potential divider junction, almost no voltage error occurs even with very high resistor values.

Any voltage between the supply voltage and zero may be set with a suitably chosen pair of resistors. The ratio of the resistor values determines the ratio of the voltages across them, as shown below.

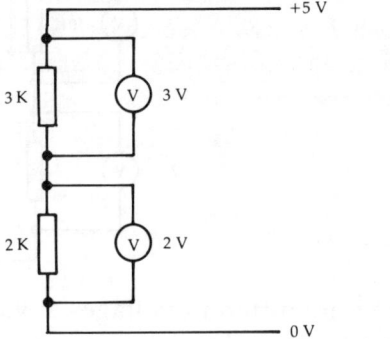
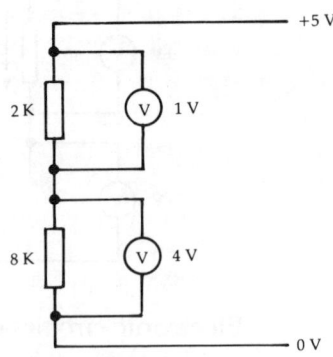

Teacher Demonstration

Resistors are inexpensive, and available from virtually all electronics suppliers. A variety of values could be purchased, and pairs joined in series by twisting (or soldering) them together. The voltage at the junction between the pairs of resistors could be measured with a voltmeter as shown below.

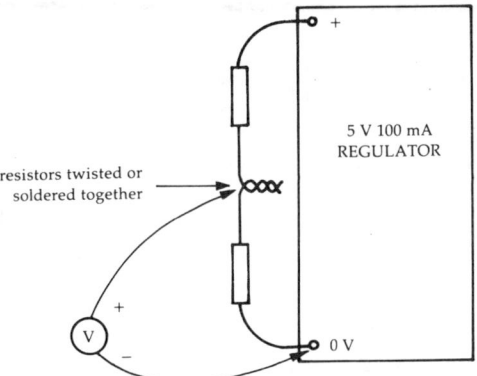

resistors twisted or
soldered together

5 V 100 mA
REGULATOR

Variable
Resistor

A variable resistor has a control knob or similar device to enable its resistance to be reduced from its stated value to zero ohms. It can be drawn in any of the ways shown below. Only two connecting points are required, but in practice variable resistors are nearly always three-terminal devices. This enables a variable resistor to be used as a *potentiometer*.

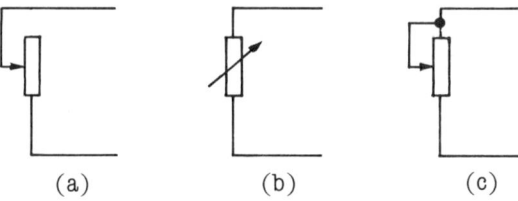

(a)   (b)   (c)

Applications

Resistors are used in abundance in electronic circuits to set voltages, control currents and 'loosely couple' one stage of a circuit to another (i.e. link one stage of a circuit to a second without a direct connection, which might allow the second stage to interfere with the first).

Variable resistors are often wired as part of a potential divider with another resistance such as a light dependent resistor. The voltage at the junction between the variable resistor and LDR is then used to control another part of the circuit (*see* Investigations 16 and 29).

Other applications include the old-fashioned stage-light dimmer units (the type which allowed the lighting technician to fry eggs and bacon on the top at the same time!).

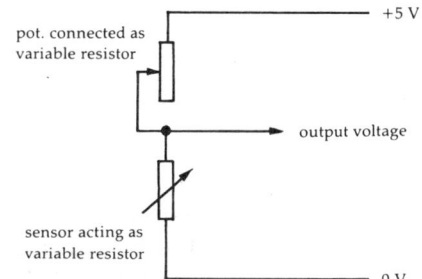

pot. connected as
variable resistor

+5 V

output voltage

sensor acting as
variable resistor

0 V

Investigation 8
Potentiometer

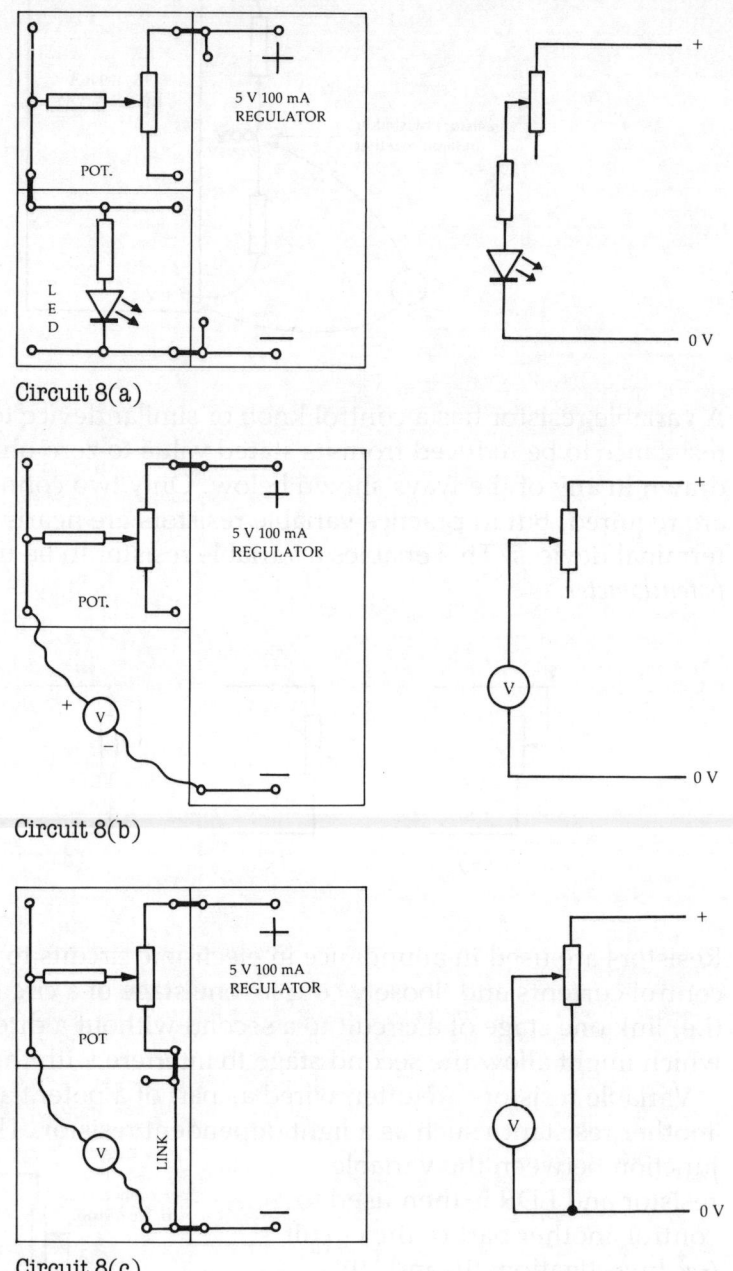

Circuit 8(a)

Circuit 8(b)

Circuit 8(c)

Circuit 8(a) employs the potentiometer (often referred to as a pot.) as a variable resistor, showing how the LED can be made brighter and dimmer by varying the total series resistance.

Circuit 8(b) demonstrates that, providing that a high resistance voltmeter is used, the flow of current through the variable resistor is too small to cause a significant voltage drop, at any setting. This should help pupils to understand that, unless a known current is flowing, the voltage across a resistor cannot be calculated.

Circuit 8(c) demonstrates that, as with the fixed resistors, a particular voltage can be created using a potential divider. The pot. is particularly suitable for this purpose, as the 'junction' between the two fixed resistors now becomes the 'wiper' of the pot. A high resistance voltmeter connected as shown in Circuit 8(c) will demonstrate the effect of the pot. when used as a potentiometer rather than a simple variable resistor. It should now be possible to set any voltage from zero to 5 volts.

Background Information

A fixed value resistor is mounted on the potentiometer board (in the Kent Electronics Kit) in order to protect the pot. against excessive current. With this series resistance, the minimum value obtainable with the pot. board is 220 Ω. Therefore the current can never be excessive. The value of the fixed resistor is small compared with the maximum value of the pot., and can normally be disregarded.

In practical circuits, when a pot. is used as a simple variable resistor, the wiper and 'spare end' are nearly always connected together as shown in diagram (c) on page 8. However, this is of little consequence in these demonstration circuits.

Further Work

A multimeter set to the resistance range (a digital type is ideal) could be connected to the pot., both to read its total resistance from end to end, and the resistance between one end and the wiper as the knob is rotated.

Homework Suggestion

Pupils could list all the uses of the pot. in their homes. They may find that they are more common than they thought.

Applications

Potentiometers are generally used when a small current flow is required, such as in modern stage-lighting controllers (where a thyristor or triac actually causes the 'dimming' rather than the pot. itself), domestic dimmer controls (again using a thyristor or triac), and in the volume and brightness controls of television sets, and hi-fi volume and tone controls. (*Note*: Thyristors and triacs are semi-conductor devices, and are similar in operation to a transistor.)

Investigation 9
Diode

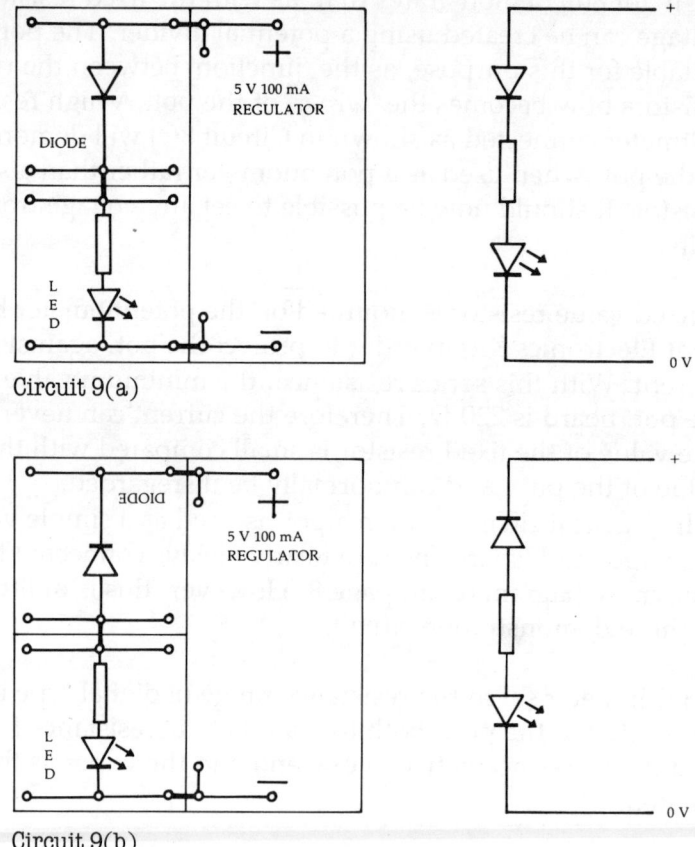

Circuit 9(a)

Circuit 9(b)

Circuits 9(a) and 9(b) demonstrate that a diode conducts in the 'forward' direction (cathode towards negative), but has a very high resistance in the reverse direction.

Circuits 9(c) to 9(g) are intended to test pupils' basic understanding of diodes, and do not have direct practical applications.

Background Information

Forward Voltage Drop

When connected in the forward direction, the diode is not a perfect conductor and may cause a potential difference between its anode and cathode. This pd can be as much as 1 V with silicon diodes (such as those used in the Kent Electronics Kit), and somewhat less in the case of germanium diodes. This can upset some experiments, particularly those in which bulbs are used.

Forward Current Rating

The type 1N4001 used in the Kent Electronics Kit has a maximum

42

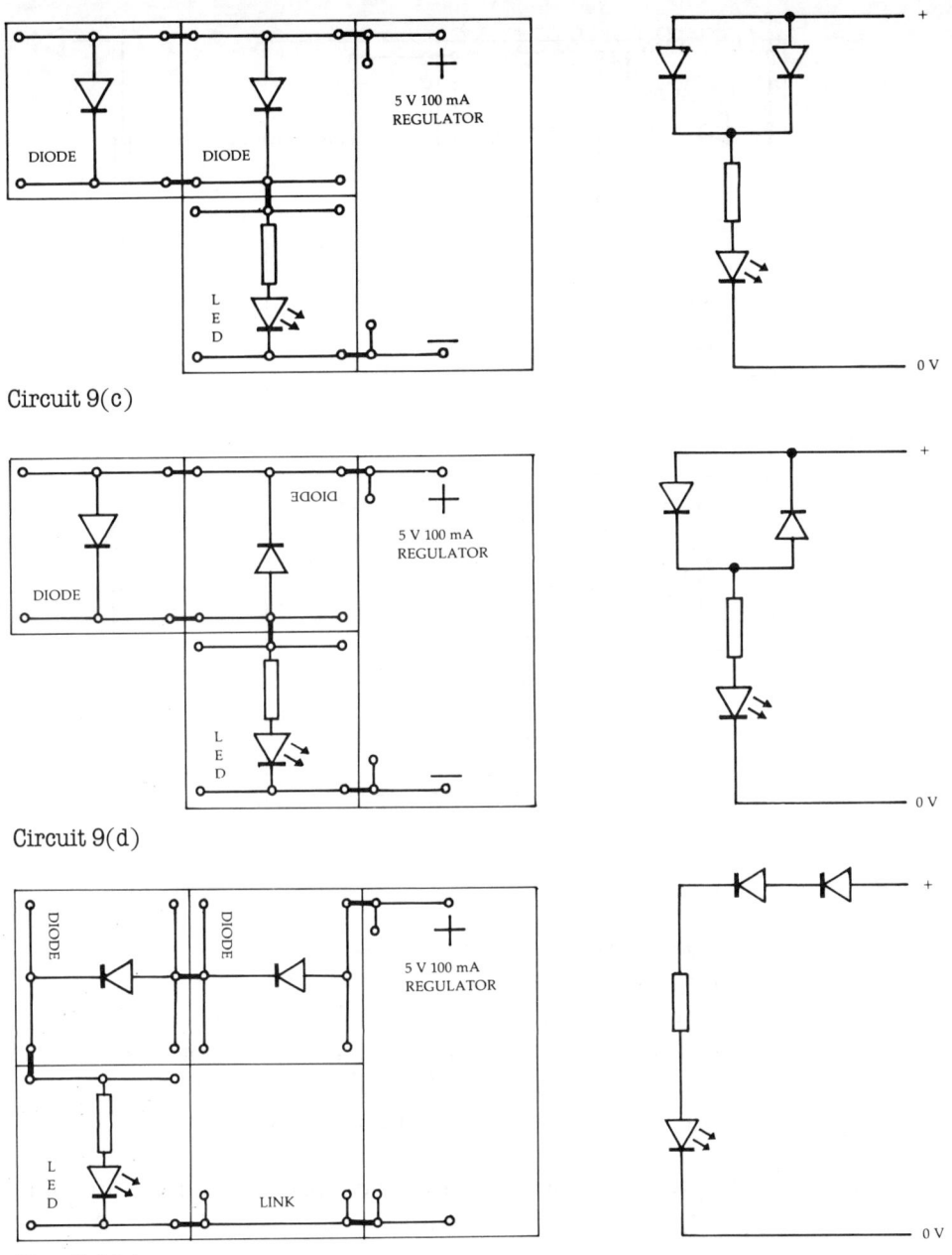

Circuit 9(c)

Circuit 9(d)

Circuit 9(e)

current rating of 1 A in the forward direction. The diode type 1N4148 (used elsewhere in the Kent Electronics Kit) has a maximum forward current rating of 75 mA. Higher forward current diodes are available for up to 25 A or more.

Reverse Leakage Current
When a reverse voltage is applied to a diode, a very small current will flow. This is known as the diode *leakage* current. It is so small that it can normally be disregarded.

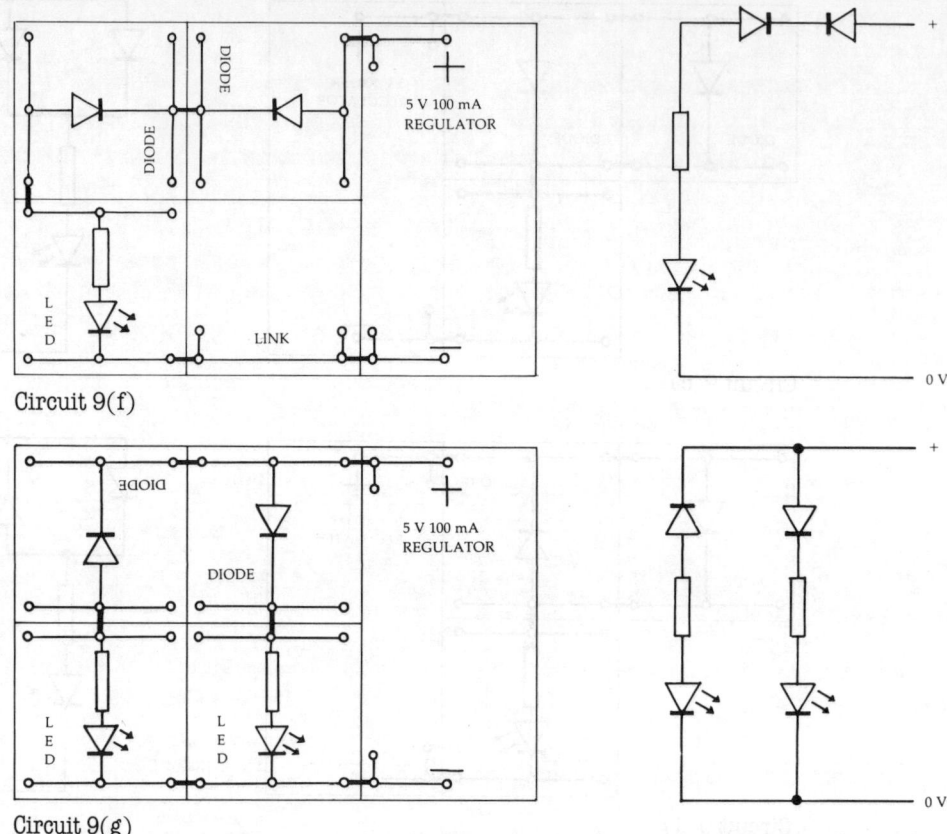

Circuit 9(f)

Circuit 9(g)

Reverse Voltage

The type 1N4001 diode has a maximum *reverse* voltage rating (known as *peak inverse voltage* or PIV) of 50 V. In other words, it will withstand a reverse voltage of up to 50 V, but beyond this an appreciable current will flow in the reverse direction. This situation is known as diode *breakdown*. Diodes with higher PIVs are available, e.g. type 1N4002 with a PIV of 100 V and type 1N4003 with a PIV of 200 V.

Further Work   An important application in logic work can be introduced by asking pupils to design a circuit with two lamps (or LEDs), A and B, and two

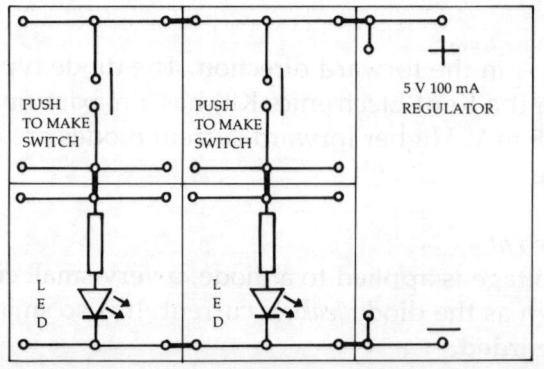

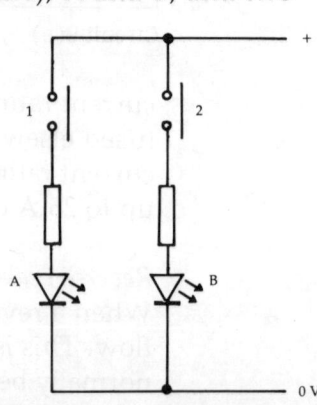

push switches, numbered 1 and 2. The circuit must be designed in order to enable switch 1 to control lamp A, and switch 2 to control lamp B. Most pupils should, hopefully, be able to tackle this.

The next problem is to modify the circuit so that switch 1 controls both lamps A and B, but switch 2 still controls lamp B only. Some pupils will link switch 1 to lamp B, and believe that they have solved the problem. Unfortunately, they will find that switch 2 now also controls both lamps, instead of lamp B only, as was required.

The problem is solved by placing a diode in the position shown below. Switch 1 will then control both lamps, but switch 2 will control only lamp B.

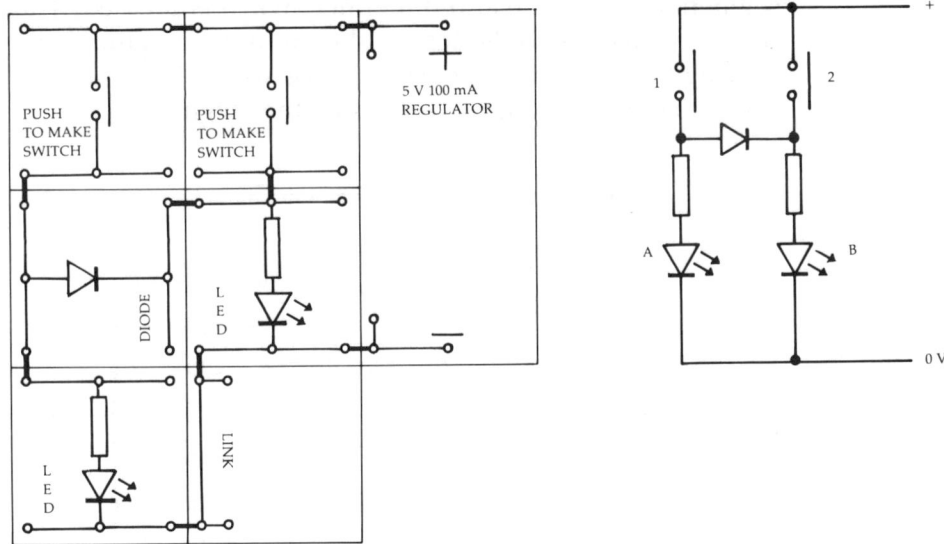

This problem can then be developed into a more practical situation involving a 7-segment display, consisting of seven LED segments, which can display any number from 0 to 9. Each segment consists of a separate LED labelled from a to g as shown below. We will consider the type known as a *common cathode* display, where all the LED cathodes are connected to a single common pin on the back of the display. Each LED anode is connected to a separate pin on the back of the display.

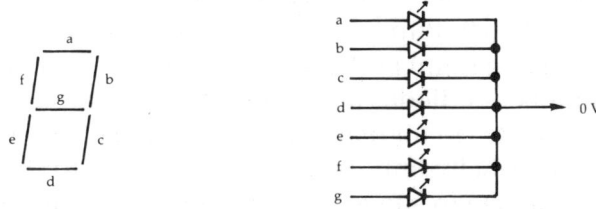

The problem can be outlined as follows: Design a circuit which consists of two push switches, numbered 1 and 2. Switch 1 must light LED segments b and c, to display the number '1'. Switch 2 must light LED segments a, b, g, e and d, to display the number '2'.

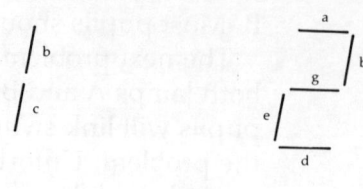

Pupils should have no problem with segments a, c, g, e and d, since they may be connected directly to the appropriate switch. However, segment b is common to both numbers '1' and '2'. When 'b' is connected to both switches, current will be able to flow to all the segments a, b, c, g, e and d if either switch is pressed. The problem is solved by using two diodes, as shown in the diagram below. Current flowing to segment b from switch 1 or switch 2 is thus prevented by the respective diode from reaching the other segments.

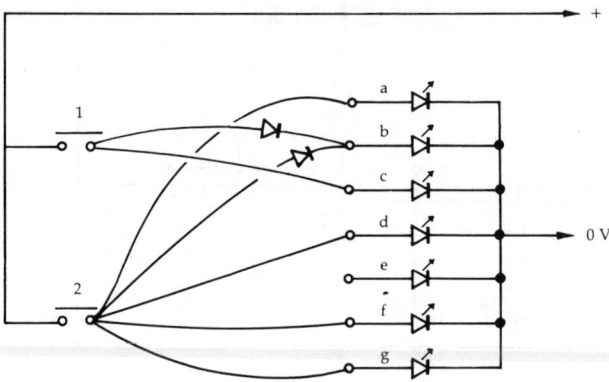

Note: In practice a series resistor must be connected to each LED, e.g.

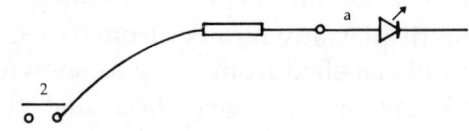

On a 5 volt supply, a value of about 330 Ω is suitable

Homework Suggestion

The problem above may be extended to cover all the other numbers from 0 to 9. Pupils will quickly find that every segment is used by more than one number (like segment b in the example above) and a large number of diodes are required. Some pupils may be able to produce a complete circuit diagram, but it is likely to be hopelessly complicated, and virtually impossible for the teacher to check. It would be an advantage to suggest that the diodes are arranged in a matrix, with a couple of example numbers provided. The complete solution is given below.

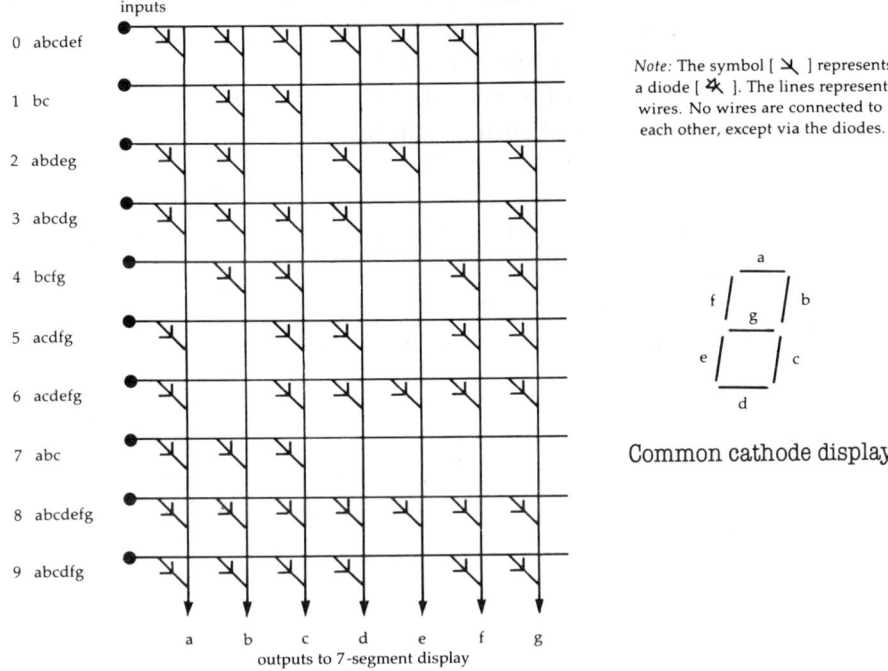

inputs

0 abcdef
1 bc
2 abdeg
3 abcdg
4 bcfg
5 acdfg
6 acdefg
7 abc
8 abcdefg
9 abcdfg

a b c d e f g

outputs to 7-segment display

Note: The symbol [⊻] represents a diode [⊷]. The lines represent wires. No wires are connected to each other, except via the diodes.

Common cathode display

In practice, the diodes would be contained in an integrated circuit, which includes other components to increase the current available at the output. Many such ICs are designed to decode binary numbers, such as the type 74HC4511 (used in the 7-segment display extension for the Kent Electronics Kit).

Further Demonstrations

Diodes are often used to rectify ac, i.e. change ac into dc. This was discussed in the section dealing with power supplies (see page 20) and a simple demonstration may be performed as shown in the diagrams below.

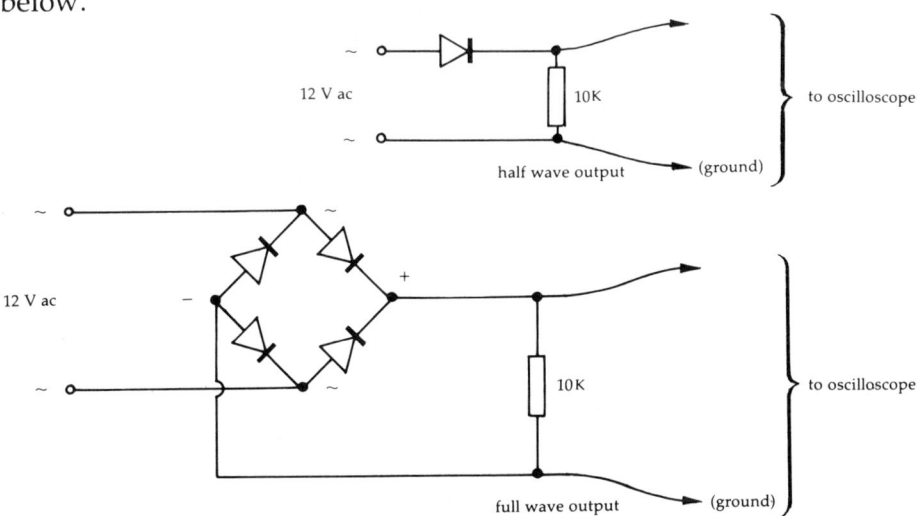

12 V ac

10K

to oscilloscope

half wave output   (ground)

12 V ac

10K

to oscilloscope

full wave output   (ground)

Great care is needed with this experiment, as the ac supply will almost certainly be unregulated. Teachers may therefore prefer to treat it as a demonstration, particularly as an oscilloscope is required.

Alternatively, a simple 'safety lead' could be manufactured. This consists of a small 470 Ω resistor fitted in series with a flying lead, as shown. The resistor could be soldered just inside the cover of the banana plug.

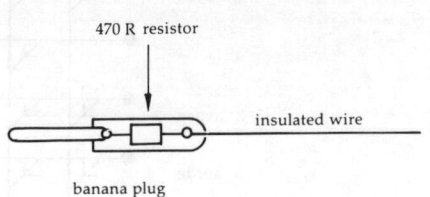

Using this lead, a pupil would be unlikely to cause damage to any part of the circuit, since even under short circuit conditions (at 12 V) only a very small current would flow. Of course, the lead would be unsuitable for general use.

Applications

Semiconductor diodes have many applications in almost every branch of electronics. They are particularly useful in logic circuits, and for changing ac to dc. They are also employed to short circuit the back emf produced by a relay (*see* page 34).

Investigation 10
Buzzer

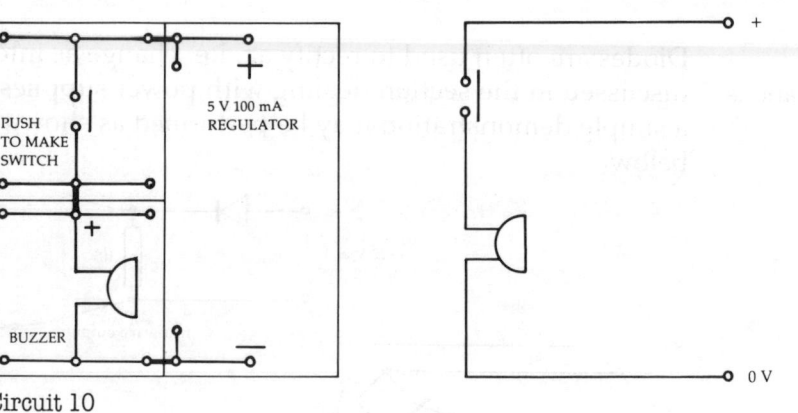

Circuit 10

The circuit shows how a buzzer may be used to convert electrical energy into sound energy, and provides an interesting alternative to the use of an LED as an output device.

Problem
Solution

The alarm circuit mentioned in the Pupil's Book causes the buzzer to sound when the magnet is placed against the reed switch. In a real-life situation, this would occur when the door or window is closed. Thus the buzzer would sound when the door is closed, and would stop sounding when the intruder entered. This is the opposite of what is required! The relay may be used to reverse the situation, by connecting its *break* contact to the buzzer as shown below.

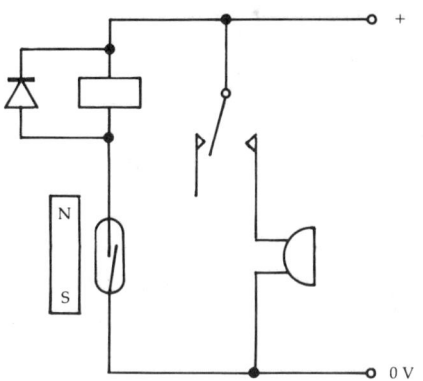

The buzzer used in the Kent Electronics Kit is a 'transistorised' or 'solid state' type. It contains a transistor oscillator circuit which produces an alternating voltage. This is converted into sound by means of a small diaphragm. The absence of any make and break contacts (as used in electromagnetic bells and buzzers) ensures that the transistorised buzzer does not produce any high voltages which might damage other sensitive electronic components.

Even so, transistorised buzzers can produce electrical 'noise' (unwanted fluctuations of voltage on the supply rails or elsewhere), and care must be exercised in their use. In most cases, problems of this kind can be overcome by the addition of a capacitor across the supply rails or, occasionally, across the buzzer itself.

It should be noted that the buzzer included in the Kent Electronics Kit is a 6 V type. It should work on the 5 V supply, providing that it is not wired in series with other components, such as diodes. When driven from a transistor, results may be disappointing because the transistor is not a perfect conductor and slightly reduces the flow of current. The problem may be overcome by driving a relay from the transistor, and switching the buzzer via the relay contacts.

Pupils could be asked to connect a buzzer and LED together in some way, so that both operate when a switch is pressed.

If the two devices are connected in parallel (observing the correct polarity), as shown at the top of page 50, they should both work correctly.

If the devices are connected in series as shown in the lower diagram on page 50, the LED should appear to glow normally, but the buzzer is unlikely to operate properly, although it may emit a weak squeaking sound! This occurs because the buzzer requires more current than the

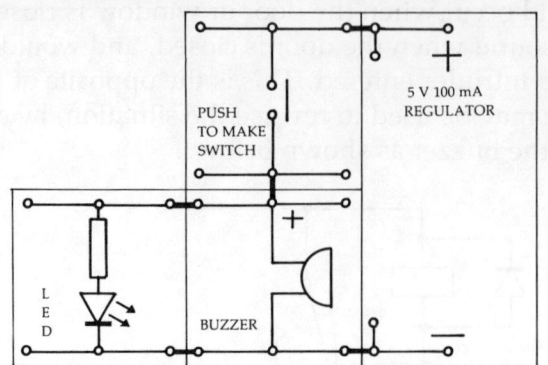
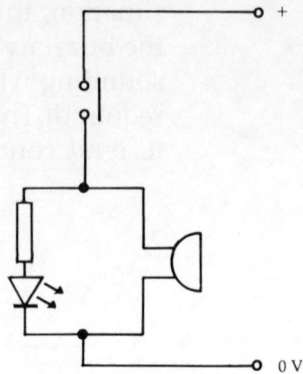

LED and resistor. The same current flows through both the buzzer and the LED. It is sufficient to light the LED, but not enough to sound the buzzer. This circuit illustrates the care needed when electronic components are connected in series.

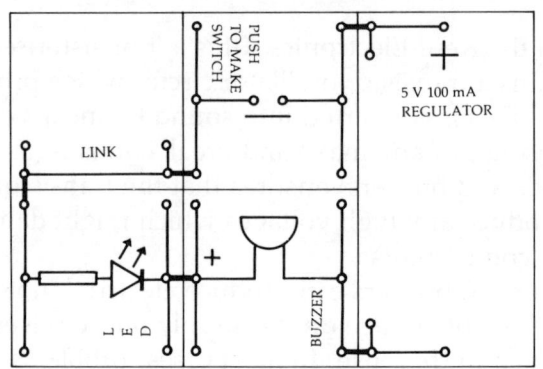
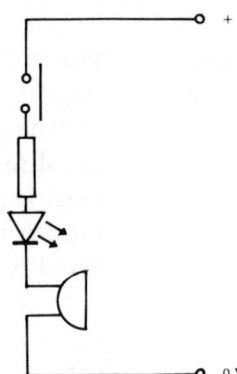

Applications

An audible warning device may be used as an alarm, for example burglar alarms, 'paper out' alarms on computer printers, alarm clocks and egg timers.

Investigation 11
Light Dependent Resistor

This investigation demonstrates that as the light incident on the light dependent resistor (LDR) increases, its resistance falls, and the LED therefore becomes brighter. This can also be illustrated by connecting a spare LDR across a multimeter set to a suitable resistance range.

Enough light should be available in a bright classroom to reduce the resistance of the LDR sufficiently to make the LED glow. In poor light conditions, a ray box or similar light source could be directed towards the LDR.

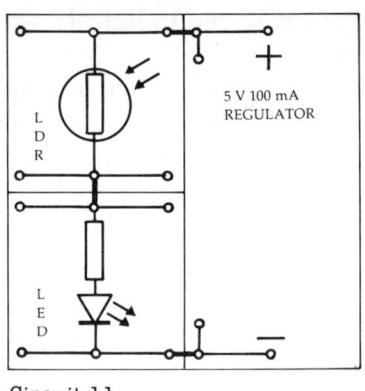

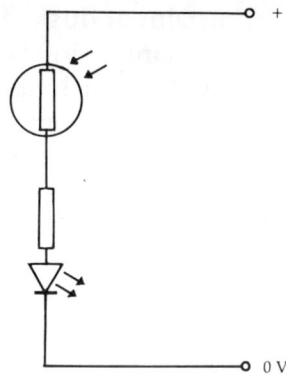

Circuit 11

Problem Solution

The LDR and the reed switch must be placed in *series* as shown below, in order to make the LED glow only when there is a bright light shining on the LDR and a magnet is placed near the reed switch.

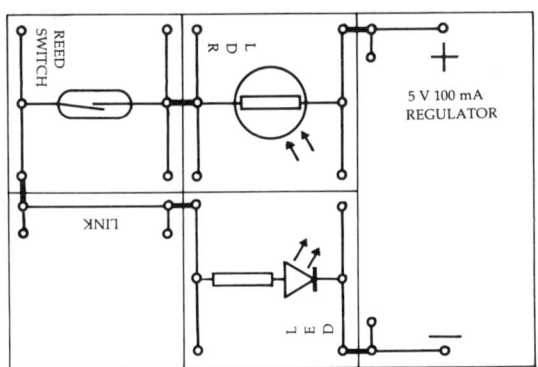

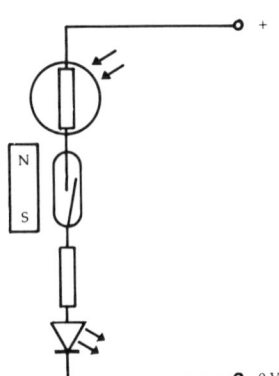

Background Information

The type of sensor used in the Kent Electronics Kit is a *cadmium sulphide cell* (CdS cell). It behaves in a similar way to a variable resistor, with a resistance range from about 100 ohms in bright light, to 10 million ohms in darkness. It should not be confused with a 'solar cell' or 'solar panel', which converts light energy into electrical energy.

Further Work

Pupils could be asked to connect the LDR directly to a multimeter set on the resistance scale. They could note the highest and lowest readings they can obtain, and perhaps compare the readings produced by different types of lamps.

Homework Suggestion

Applications of LDRs could be listed, from automatic bicycle lights to counting objects on a conveyor belt as they break a beam of light.

Applications

Light dependent resistors are used extensively in light sensitive switches, for example devices for switching on lamps or drawing

51

curtains at dusk. They are also employed in light-beam detector circuits, for example in alarm systems.

Other much faster devices are available, such as *phototransistors* and *photodiodes*. Some are designed for use with infra-red light, and can be used for measuring speed, counting objects, and in optical tranmission systems such as fibre-optics.

Further applications of LDRs are listed in Investigation 16 (*see* page 68).

Investigation 12
Thermistor

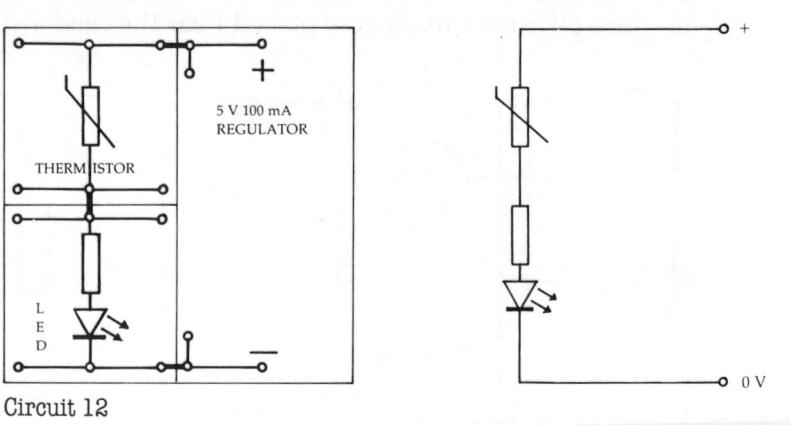

Circuit 12

When the thermistor is warmed up by placing a finger against it, the LED (in theory) will glow more brightly. This shows that the thermistor is a temperature sensitive device, whose resistance falls as its temperature rises.

The change in resistance is not as great as that of the LDR in Investigation 11, and the change in brightness of the LED may be hardly noticeable. A much hotter object placed against the thermistor may produce better results, but *do not use* a naked flame.

While this particular experiment (with the LED) may appear disappointing to pupils, the lack of a good result can be used to advantage when a transistor or IC is employed later. The excellent results achieved when a transistor and/or IC is used with the thermistor emphasise the usefulness of devices which provide voltage and/or current gain.

Further Work   A multimeter set to a suitable resistance range may be employed to demonstrate the change in resistance of the thermistor. Pupils could then compare the readings obtained with the thermistor placed in a fridge, or near a heater. A spare thermistor could be mounted at the end of a glass tube. If sealed properly, this could then be used as a probe to compare the temperature of different liquids.

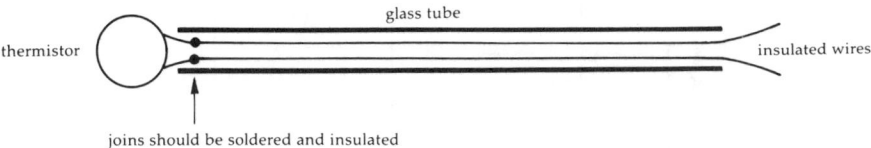

joins should be soldered and insulated

Applications   A temperature sensor may form the heart of a thermostat to control a central heating system, or a fridge or freezer. Other uses include greenhouse high/low temperature warning systems, and a road-ice alarm on a car.

Investigation 13
Capacitor

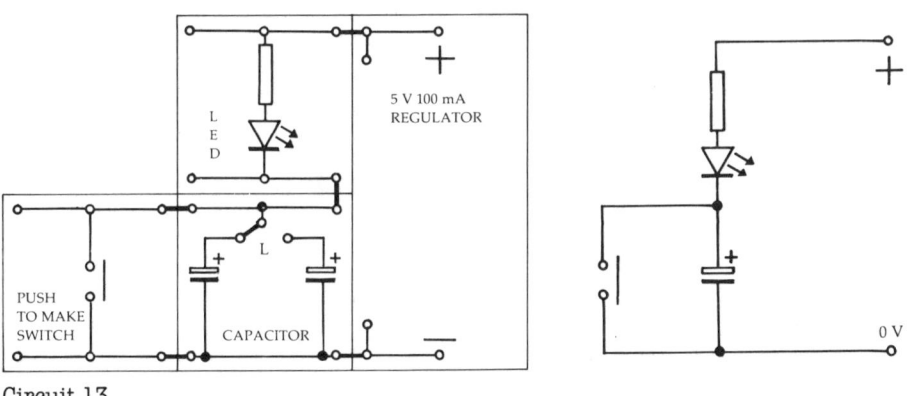

Circuit 13

Circuit 13 demonstrates that when a capacitor is charged via a series resistance (the LED and associated resistor in this case), it takes a certain time to fully charge. This charging time is indicated by the time for which the LED glows. Pressing the push switch 'shorts out' the charge almost instantly, since there is no limiting resistance other than the resistance of the printed circuit board tracks, switch contacts etc. When the switch is released, the capacitor charges once more via the LED.

Two sizes of capacitor are available, the smaller one charging up in less time than the larger type. Thus the circuit represents a primitive fixed value timer. The time may be increased by increasing the series resistance. This can be achieved by placing the pot. (connected as a variable resistor) in series with the LED board. The charging time can now be substantially increased, but the LED may be dim, due to the small current.

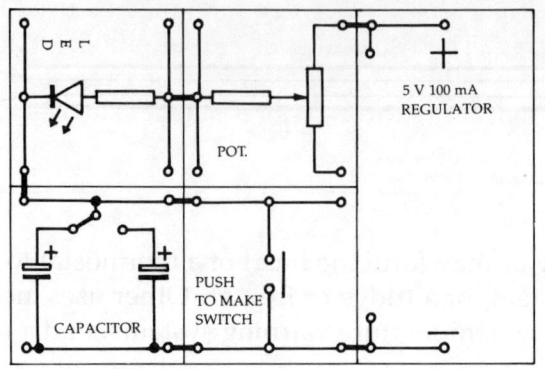 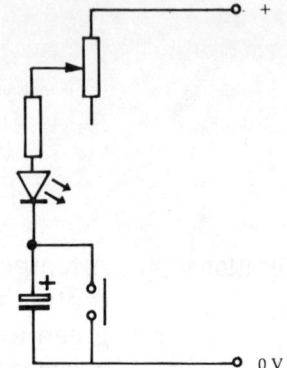

Further Work   When the capacitor is fully charged (i.e. the LED has stopped glowing), it will retain its charge for some time, providing that the push switch is *not* pressed. This can be demonstrated by carefully removing the capacitor board from the circuit, and connecting the LED board as shown in the diagram below.

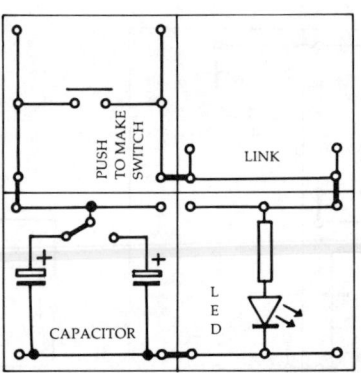 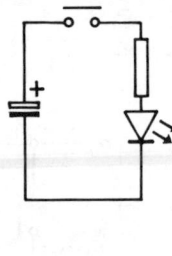

When the switch is pressed, the LED should glow for a short time as the capacitor discharges. The experiment may be repeated providing that the capacitor is charged up again by, for example, connecting it across the output from the 5 V regulator.

'Special Effects'   Spectacular results may be obtained if a large value, e.g. 4700 μF (or greater) capacitor, rated at 25 V or more, is obtained. This capacitor may be connected directly across the output of a power unit set to about 20 V dc. It is important to observe the correct polarity, noting that the capacitor *negative* is normally indicated as such, and that the positive end is often coloured *black* (*note:* black not red). A spark will probably occur as the wires from the power unit are placed against the leads of the capacitor. Charging will take less than a second.

The power unit may then be disconnected from the capacitor, and a wire used to 'short out' the two capacitor leads. As the capacitor discharges, a large spark should occur.

An oscilloscope (with a very slow sweep rate) may be used to show the capacitor charging. The oscilloscope leads should be placed across the capacitor (observing the correct polarity). The capacitor should be made to charge or discharge through a resistance, such as the pot., as shown below.

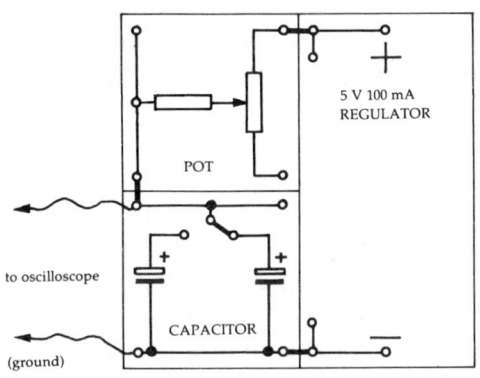 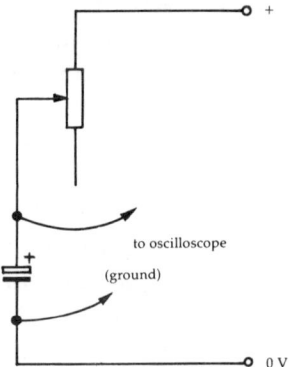

Set the pot. to about mid-way to begin with, and select the smaller capacitor. Wait for the spot to enter the left-hand side of the screen, then switch on the power supply to commence charging. The result should be similar to the graph shown on the right.

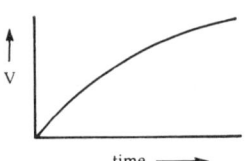

Experiment with different settings of the pot., oscilloscope sweep rate and capacitor size in order to see the effect. Turn off the power supply and short-circuit the capacitor with a flying lead before repeating the experiment.

Capacitor discharging may also be shown by an oscilloscope. A suitable circuit is given below.

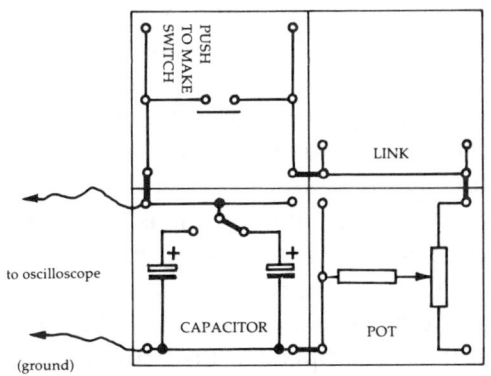 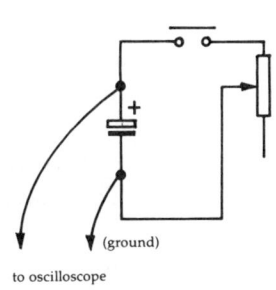

Press the push switch when the oscilloscope spot enters the left-hand side of the screen. Experiment with the values and settings as before, in order to find the best effect. The graph shown on the right should be obtained on the oscilloscope screen.

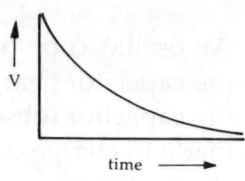

The capacitor may be re-charged by briefly connecting it via flying leads to the output of the regulator, ensuring that the correct polarity is observed.

Charging 'Time Constant'

The capacitor charging rate becomes slower as the voltage across the capacitor increases and, in theory at least, the voltage across the capacitor will never reach the supply voltage. However, the voltage across the capacitor will definitely reach two-thirds of the supply voltage, and the time taken for this to occur is called the *time constant* of the circuit. This is given by the formula:

$$T = R \times C$$

where T is the time constant in seconds, R the series resistance in ohms, and C the capacitance in farads. This formula is particularly useful when using capacitors with the CMOS logic ICs described later, since the input 'trigger' voltage at which the IC output changes state is about two thirds of the supply voltage.

Homework Suggestion

If the charging and discharging graphs have been demonstrated, pupils could be asked to sketch some graphs obtained using several sizes of capacitor and values of resistor. A *higher* value capacitor and/or *higher* value resistor will produce a *slower* charge/discharge rate.

Applications

Capacitors are frequently used in timing circuits, for periods from fractions of a second to many hours. Capacitors are often employed to control the frequency of an oscillator, for example in an electronic organ. In a similar way, capacitors (often variable types) are used as part of a tuning circuit in radio receivers and transmitters.

Large value capacitors are used in smoothing circuits, to convert a varying dc supply into a supply suitable for electronic circuits, as outlined on page 20. Small value capacitors are used to remove voltage spikes on power supply lines, and in general 'decoupling' applications (to prevent the supply voltage fluctuating as the current changes).

The plates of a capacitor are insulated from each other (*see* page 10). Thus dc is unable to flow through a capacitor. However, a changing voltage on one plate will cause a change of voltage on the other plate. An ac supply will therefore appear to flow through a capacitor.

The ability of a capacitor to block a dc supply, but conduct ac makes it useful in 'coupling' applications where two sections of a circuit (e.g. an amplifier) can be coupled (linked) together. The capacitor allows the ac (audio) signal to pass from one stage to another, but prevents the passage of dc, which would otherwise upset the other dc voltages in the two sections of the circuit.

INTRODUCTION TO TRANSISTOR CIRCUITS

The transistor is a very important and fundamental device in electronics. Essentially, the transistor is like a variable resistor, whose resistance is controlled by its input voltage. Since small electrical signals at the input can cause much larger electrical signals at the output, the transistor is often used as an *amplifier*. A transistor may also be employed as an electronic *switch*, and this application is explored in the Kit. Using a transistor as a switch greatly increases the scope and usefulness of the sensors and components already described.

The transistor used here is a 'bipolar' *n-p-n* type. It has three pins or legs known as the base, collector, and emitter. The *p-n-p* type transistor is not used in the Kit, but has similar properties except that the polarity is reversed.

The following notes assume that the negative supply rail is called the 'zero volts rail'.

Symbol
for an *n-p-n*
transistor

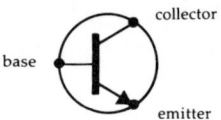

There are a number of possible methods of connecting a transistor in a circuit, but the transistor supplied with the Kent Electronics Kit is connected in 'common-emitter mode', where the emitter is connected to the zero volts rail, which is 'common' to both the input circuit and the output circuit.

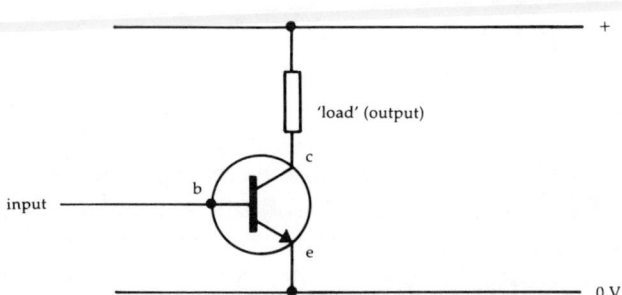

Base

This is the input to the transistor. The base/emitter junction inside the transistor is like a diode with a forward voltage drop of 0.6 V. If a voltage greater than 0.6 V is applied across the base and emitter, current will flow from the base to the emitter. This current causes a much larger current to flow from the collector to the emitter, assuming that the collector is at a higher potential than the base.

Collector

The collector pin is usually connected to the positive rail through a resistance (e.g. resistor, bulb, buzzer etc.) called the *load*.

Emitter	The emitter is shown connected to the zero volts rail, and completes the circuit from the positive rail to the zero volts rail.

Switching on an *n-p-n* transistor

The following applies only to an *n-p-n* transistor and assumes that all voltages are measured relative to the zero volts rail, to which the emitter has been connected. When a small voltage of less than 0.5 V is applied to the base of a transistor, the transistor remains switched *off*. In other words, there is a very high resistance between the collector and emitter connections, and virtually no current can flow from collector to emitter.

If the base voltage is increased to 0.6 V, the transistor begins to switch *on*, and at about 0.75 V the transistor is switched fully on and is said to be *saturated*. The resistance between the collector and emitter connections is now quite low, and current is able to flow from positive through the load to zero volts via the transistor.

Transistor as a Switch

The transistor can be used as a high speed switch, by changing the base voltage between 0.5 and 0.8 V. This can be achieved in various ways; however, many of the circuits in this Guide use a variable potential divider arrangement to adjust the voltage at the base of the transistor.

Input Devices

The following are examples of input devices for a transistor:

1 Light dependent resistor (LDR) — sensitive to light
2 Thermistor — sensitive to temperature
3 Switch (eg. push to make) — sensitive to pressure or movement

4 Capacitor/resistor — provides a time delay
5 Reed switch — sensitive to magnetism
6 Microphone — sensitive to sound

The devices listed above communicate information to the transistor. The transistor will act upon this signal, perhaps by switching on if the conditions are right. This information could then be sent to an output device.

Output Devices

The following can be used as output devices for a transistor:

1 Filament bulb   4 Relay
2 Light emitting diode (LED) 5 Solenoid
3 Buzzer   6 Loudspeaker

The transistor employed in this Kit is a type BC184L. It has a gain of more than 200 times (known as its h_{FE}). For example, a current of 1 mA into the base of the transistor could cause a current of about 200 mA to flow between the collector and emitter. The actual current flowing will depend on other factors including the load resistance and voltage. The

BC184L has a maximum permissible collector/emitter current of 200 mA and will operate on a supply of up to 30 V.

Since the base/emitter junction is like a forward biased diode, if the base were connected directly to the positive supply, a very large current might flow. A resistor is therefore required to limit the current to a safe level in order to protect the transistor. In this Kit, a base resistor of 2.2 kΩ is included on the transistor board.

Investigation 14
Transistor as a Switch

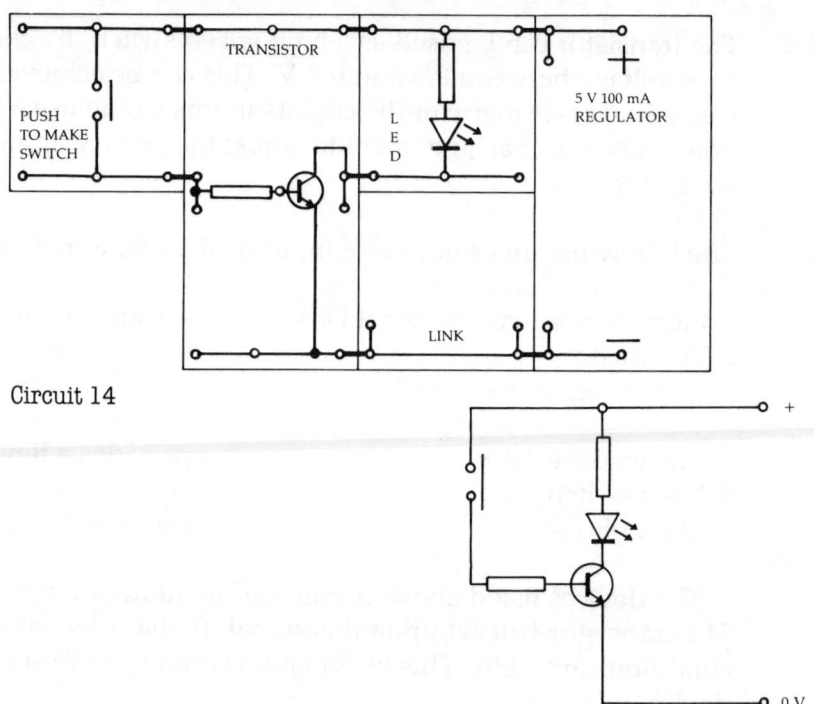

Circuit 14

The circuit demonstrates how a transistor may be switched on by connecting its base to the positive supply via a resistor and push switch. The resistor prevents too much current flowing into the base. At first sight, the circuit may not appear particularly useful to the pupils, but it should be noted that a very small current into the base switches on a much larger current via the collector/emitter leads.

When moist fingers are placed across the push switch, the moisture should conduct sufficient current to switch on the transistor. The usefulness of the transistor should now become more apparent, since moist fingers placed across the push switch in Circuit 2(a) did *not* cause

60

the LED to glow, as the moisture does not conduct sufficient current to light the LED directly.

Further Work   More experienced pupils could insert an ammeter (fsd about 10 mA) in place of the link connecting the base to the push switch. A further ammeter (fsd about 100 mA) could also be connected in place of the link connecting the collector to the LED. This arrangement is shown opposite, and illustrates that the transistor achieves a gain, i.e. it amplifies.

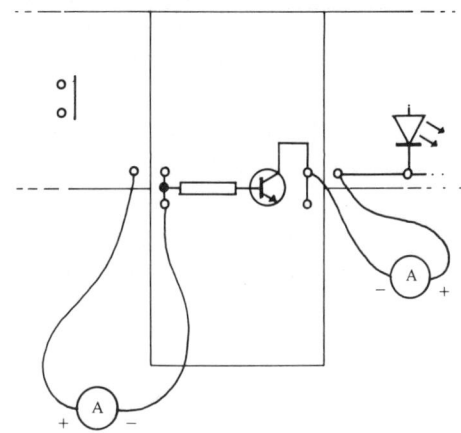

Rain Warning Alarm   Further experiments would benefit from the use of a pair of flexible leads connected (e.g. via crocodile clips) to the sockets above and below the push switch. The other ends of the leads could now enable the circuit to detect water, for example in a bath level indicator, or a windscreen-wash level indicator. A device suitable for detecting rain could be made from two strips of metal foil stuck to a piece of wood or plastic as shown below.

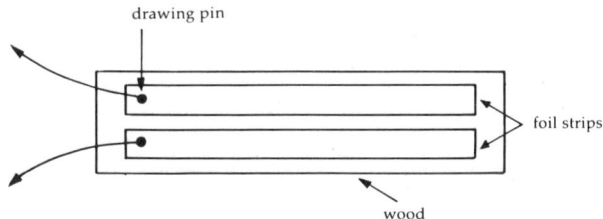

If the device is placed outside, any rain falling between the foil strips would cause the circuit to switch on. A similar device may be produced using a small piece of stripboard (often called 'Veroboard'). Several alternate tracks could be connected together with pieces of wire to produce a larger area for rain detection.

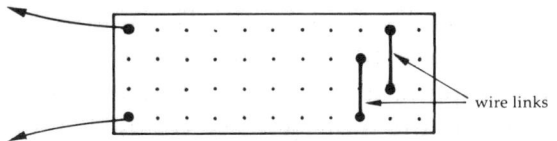

A rain warning device would be of much more use if the LED were to be replaced by a buzzer. However, in practice some buzzers may fail to work properly, as the transistor is not a perfect conductor even when saturated (turned on fully). Pupils who wish to use the ideas learned from this investigation to make a 'rain warning alarm' should obtain good results with the buzzer, if the circuit is used on a 9 V supply, e.g. a PP3 battery. Alternatively, a relay could be used between the transistor and buzzer to allow operation on 5 V, as shown in Investigation 19.

Homework
Suggestion
Pupils could be asked to design a complete water alarm circuit with sensors below all the water tanks and sinks in the house, and at other 'danger spots', for example the washing machine. They should decide whether the sensors should be wired in series or parallel. A possible solution is given below.

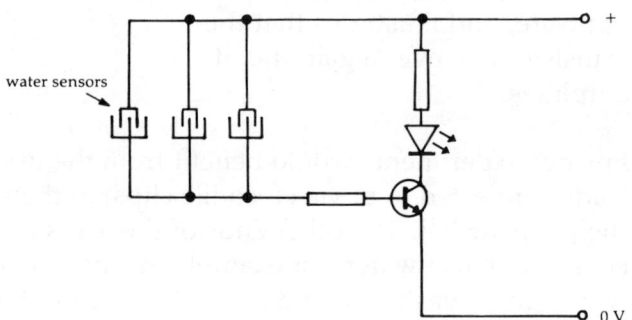

Note that the sensors are wired in parallel with each other. Connecting the sensors in series would only work if the house was completely flooded, from the ground floor to the loft!

Applications
Possible applications of this circuit include a moisture detector for potted plants, rain warning alarm, bath level indicator, 'cups full' indicator (for blind people) and 'nappy wet' indicator.

Investigation 15
'Switch on' Voltage for a Transistor

This circuit allows the base voltage of the transistor to be controlled by means of the potentiometer. The approximate point at which the transistor switches on is indicated by the LED beginning to glow.

It is important to note that the base voltage cannot exceed about 0.75 V, i.e. the transistor will conduct sufficient current from the base to emitter to prevent the voltage at the base rising much beyond this value. For example, if the voltage at the wiper of the pot. rises to 2 V, the base series resistor (fitted on the transistor board) will create a potential difference of about 1.25 V.

The transistor should start to switch on when the base voltage

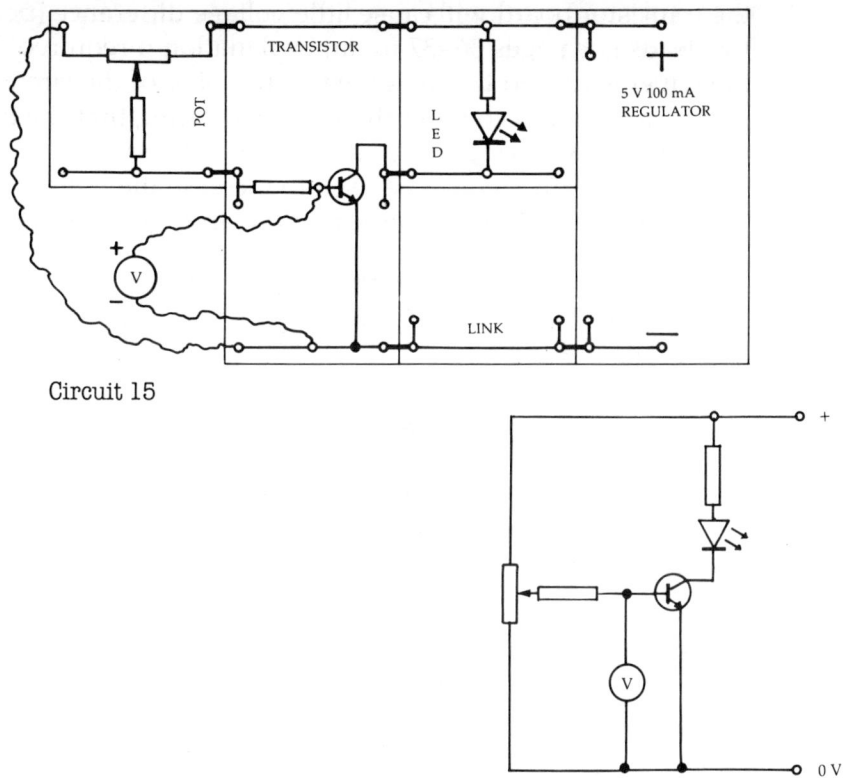

Circuit 15

reaches a little over 0.5 V, and is switched on fully (saturated) when the base voltage reaches about 0.75 V. In order to switch on, a transistor requires sufficient current flowing into the base to cause the required current to flow through the output load.

Further Work

The voltage on the wiper of the pot. could be measured by connecting the positive lead of a voltmeter to the spare socket at the left of the transistor board, as shown in the diagram below.

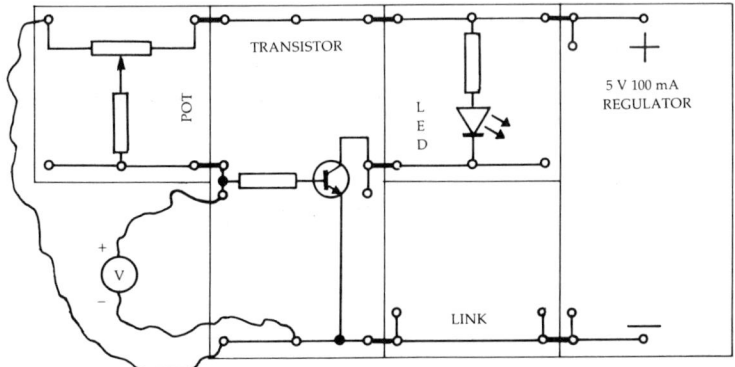

The 'switch on voltage' should have almost the same value as it did before, since very little current will be flowing, and the resistor fitted to the transistor board will cause little voltage difference (pd) between its two leads (*see* pages 36–37 for an explanation if required). However, as the wiper is turned towards the positive side of the circuit, the voltage registered will rise to more than 4 V, indicating that the base series resistor is now having a substantial protective effect.

A potential divider can be used to increase the voltage at which the transistor switches on, as shown in the arrangement below.

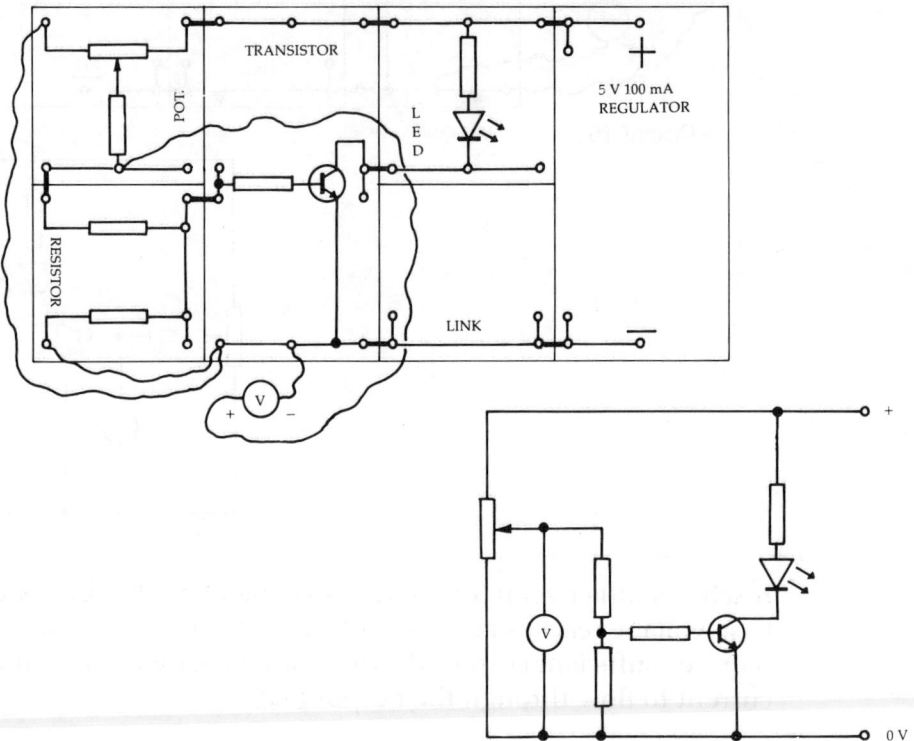

The actual switch on voltage of the transistor is still about 0.6 V at the base, but with the additional two 10 kΩ resistors, the voltage required at the 'input' to the potential divider is just over 1 V. This arrangement is useful when the device feeding the transistor base is incapable of producing a voltage lower than 0.6 V. The following circuit would switch on the transistor when the potential divider input voltage reached 2 V.

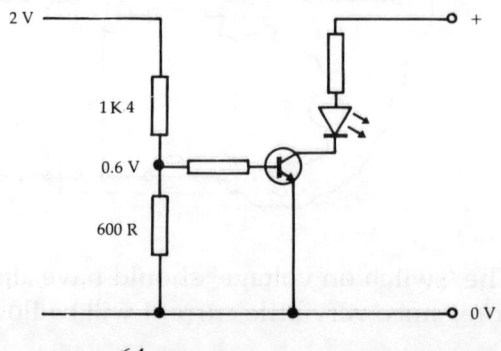

Pupils could design a transistor circuit which will *just* switch on the LED when the pot. wiper is set exactly mid-way between the two ends. They could then calculate the values of the two resistors which form the potential divider. Pupils should be asked to disregard the base current flowing (i.e. assume that the base current is zero). The solution is given below.

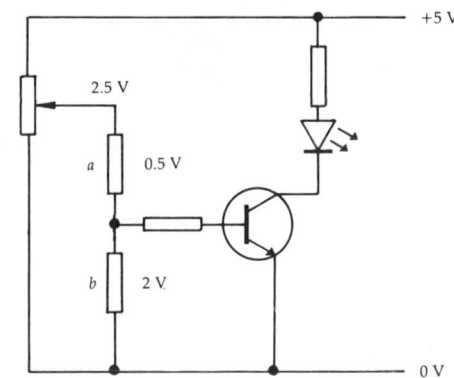

The two resistors may be of any values, providing that the *ratio* of resistances b to a is 0.5 to 2, i.e. 1 to 4. More advanced pupils could test their results with individual resistors (if available). A value of 1.2 kΩ for resistor b and 4.7 kΩ for resistor a should provide good results if a high resistance digital voltmeter is used.

Many of the circuits which follow employ a potential divider (often the pot. wired as a variable resistor in series with another resistance, such as the thermistor). When set up correctly, the voltage at the junction between the two resistances is just above or just below the transistor switch on voltage. In this case the transistor acts as a voltage-controlled switch.

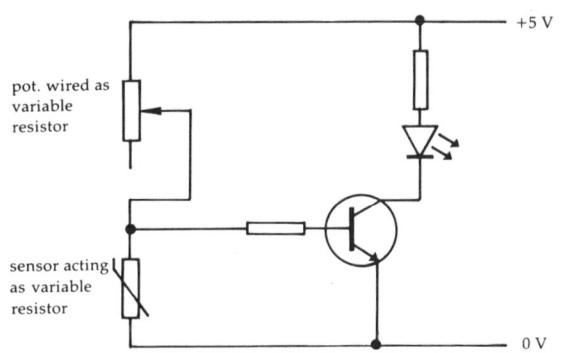

Investigation 16
Light Sensor

Circuit 16(a) shows how the LED can be made to glow when the light falling on the LDR decreases.

The pot. is connected as a variable resistor, and is placed in series with the LDR. The resistance of the LDR decreases as the light level rises. The variable resistor and LDR therefore form a potential divider with their junction connected to the base of the transistor. The voltage at this junction will depend upon the resistances of the pot. and LDR.

In practice, the pot. is set so that the voltage at the transistor base is just under 0.6 V. This ensures that the transistor is switched off. When the LDR is shaded, its resistance rises. Hence the voltage at the junction rises, causing the transistor to switch on.

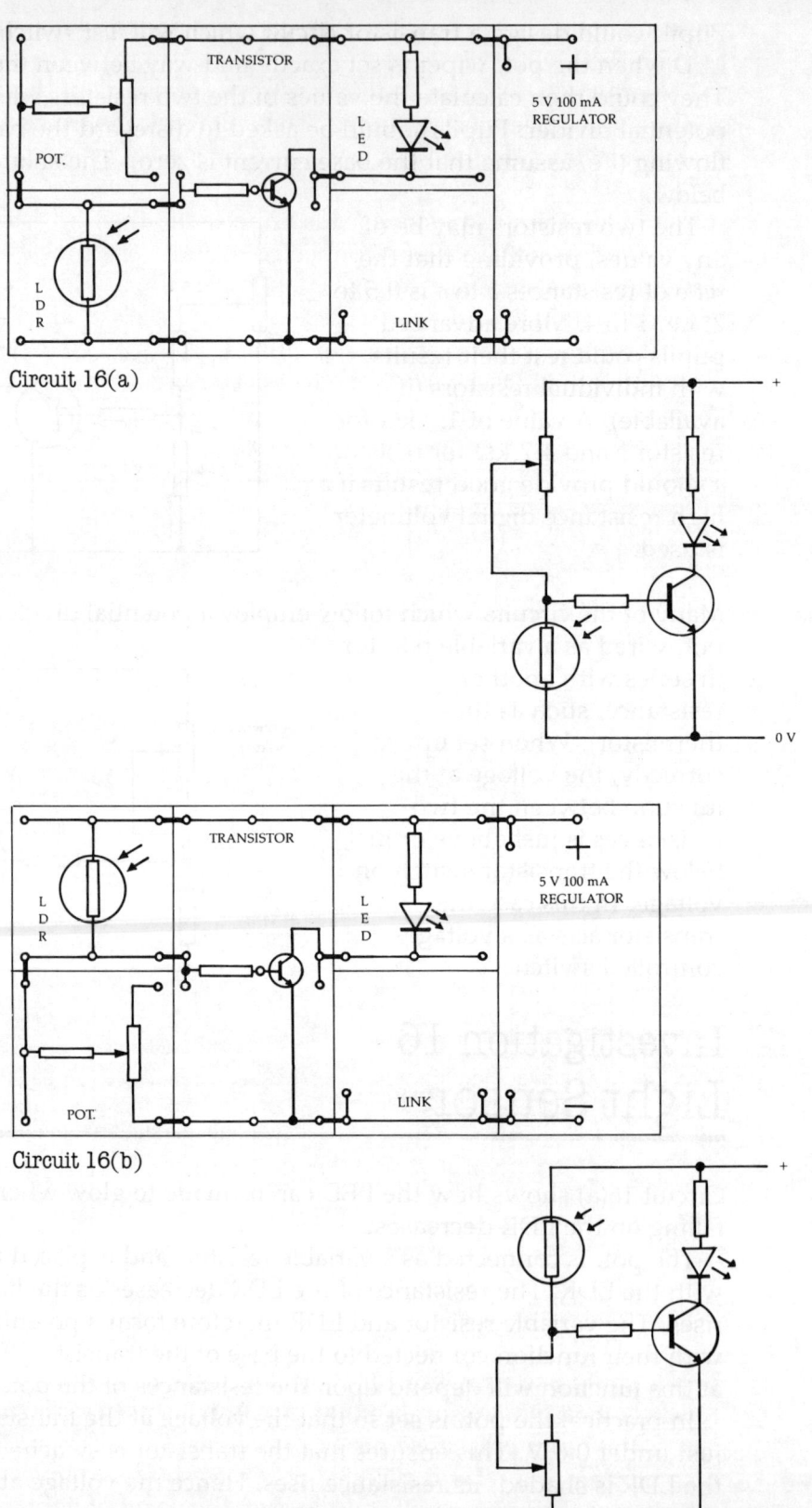

Circuit 16(a)

Circuit 16(b)

Circuit 16(b) shows how the LED can be made to switch on when the light on the LDR *increases*.

The principle is similar to that in Circuit 16(a), except that the LDR and pot. have been interchanged. The voltage at the transistor base therefore *falls* when the LDR is shaded.

Further Work

A voltmeter, with its positive lead connected to the transistor base socket, and its negative lead connected to zero volts (negative), will show the transistor base voltage changing as the LDR is shaded or exposed to light.

Further practical applications of either circuit may be explored by connecting the LDR board to the rest of the circuit via flying leads, as shown below.

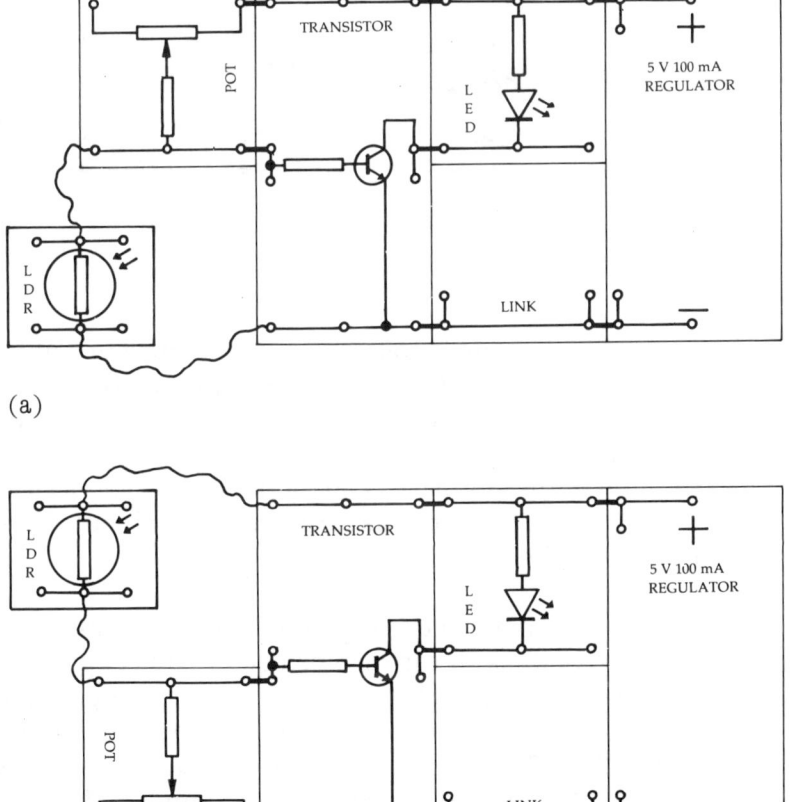

(a)

(b)

The LDR board may then be placed in any position. For example, pupils could use a ray box (or projector) to direct a beam of light towards the LDR. Assuming that Circuit (a) is used, a person walking across the beam will cause the LED to glow. The LED could be replaced by a buzzer, or a relay driving a buzzer, as shown in Investigation 19, in which case the buzzer will sound when the beam of light is broken.

Clearly, this circuit can be used as an intruder alarm. Some pupils may be able to discover how to make the relay latch on (*see* below).

If Circuit (b) is made to latch on when triggered, it could be used to switch on a light in a darkened room when a torch is aimed at the LDR.

Homework
Suggestion

Pupils could be asked to design a circuit based on Circuit (a) above to detect an intruder. The circuit should latch on when triggered. A possible solution is given below.

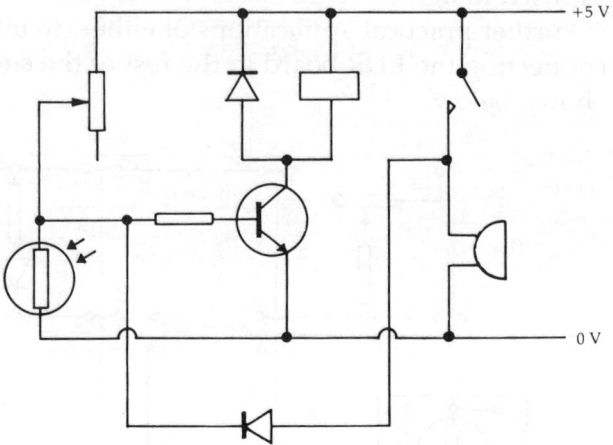

The circuit is made to latch on by feeding current from the relay output, back to the transistor base input. A flying lead could be tried (without the diode shown). The circuit may work, but will be relatively insensitive to changes in light, as the buzzer will conduct current away from the transistor base. The diode may be connected via two flying leads in order to allow the 'feedback action', but prevent current flowing in the opposite direction.

Other
Applications

Circuit 16(a) may be employed to switch on lamps, or motors (to draw curtains) at dusk. It can also be used to detect persons or objects as described above, to trigger an alarm, open a door, or activate a counting circuit, for example to count objects on a conveyor belt. The LDR is a CdS cell (*see* Investigation 11) and does not react quickly to changes in light level. It should not therefore be employed to count objects moving at high speed.

Circuit 16(b) could be used to switch on a porch light, or activate a motor to open a garage door on receiving a flash of light from a car's headlamps. Infra-red sensing devices are frequently used to control televisions and other appliances from a hand held infra-red transmitter. The infra-red receiver is normally a type of photodiode, which reacts much more quickly that an LDR. In these devices infra-red signals are coded to prevent accidental triggering from other devices producing infra-red light.

Investigation 17
Temperature Sensor

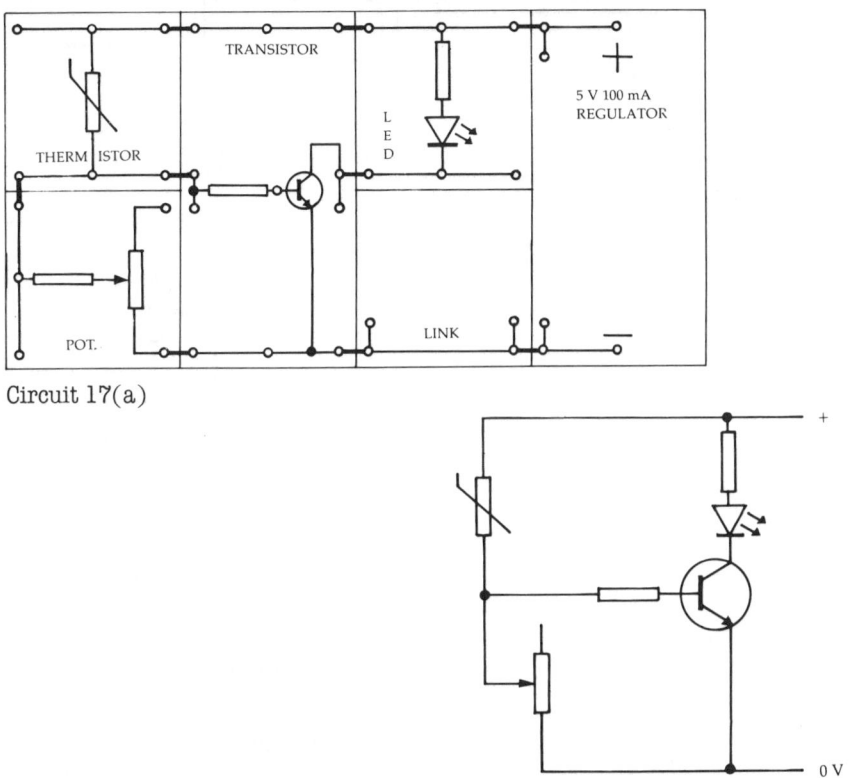

Circuit 17(a)

Circuit 17(a) demonstrates that as the temperature of the thermistor rises, the LED glows more brightly. It may be necessary to carefully set the pot. so that the LED only just glows *before* the thermistor is touched. A slight increase in the brightness of the LED should then be observed when the thermistor is touched between the fingers. Pupils who may have been disappointed by the previous thermistor experiment (Investigation 12, which placed it in series with an LED) should note the usefulness of the transistor in increasing the effect.

A thermistor behaves like a variable resistor, whose resistance depends on its temperature. The pot. is connected as a second variable resistor. This is connected in series with the thermistor to form a potential divider, with the junction between them connected to the transistor base.

When the pot. is set correctly, the voltage at the transistor base will be just under 0.6 V, and the transistor will therefore be switched off. If the temperature of the thermistor rises, its resistance will fall, and the voltage at the junction (and thus the transistor base) will rise. The transistor will switch on, allowing current to flow via the LED.

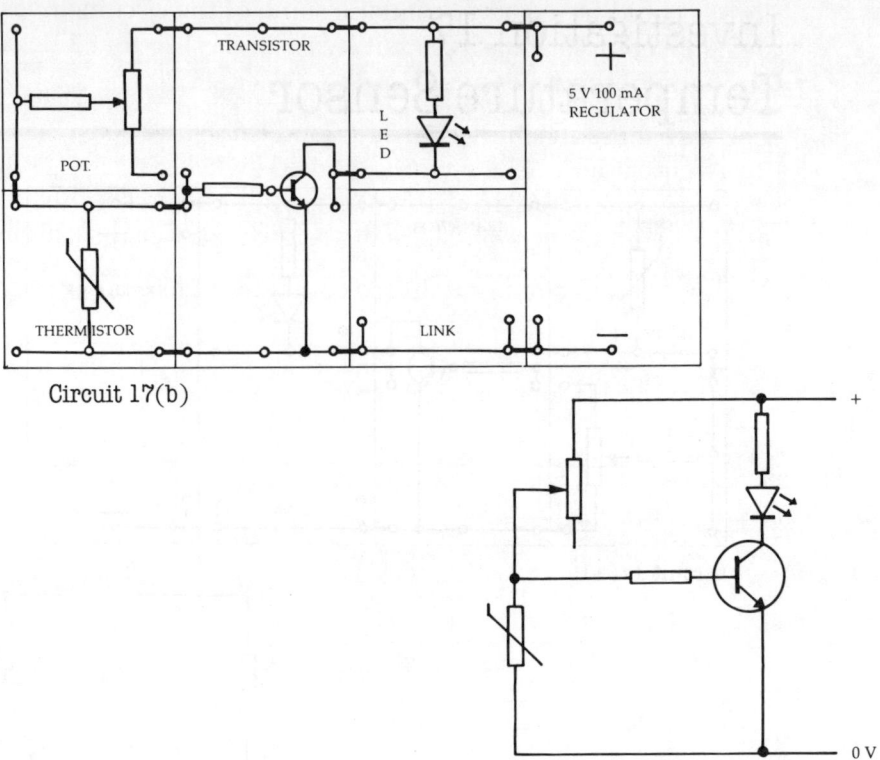

Circuit 17(b)

Circuit 17(b) demonstrates the opposite effect. This time the pot. should be set so that the LED is a little less than fully bright. It should then dim when the thermistor is touched.

The thermistor and pot. have been interchanged in this arrangement. The theory is similar to that of Circuit 17(a), except that when the resistance of the thermistor falls, the voltage at the junction, and hence the voltage at the base of the transistor, also *falls*.

When the pot. is set correctly, the voltage at the transistor base will be about 0.7 V. This is sufficient to switch on the transistor. As the temperature of the thermistor rises, the base voltage falls to less than 0.6 V, switching off the transistor and LED.

Further Work

A voltmeter may be connected, as in the previous investigation, to measure the transistor base voltage during the experiment.

Like the LDR, the thermistor may be connected via flying leads if required. Alternatively, a spare thermistor could be fixed at the end of a glass tube, and sealed to prevent water entering the tube (see page 53).

The thermistor could then be used as a probe to compare the temperatures of different liquids, by noting the position of the pot. at which the LED just begins to glow.

Pupils could also be asked to design an 'ice alarm'. They should check that when the thermistor is placed in ice, the LED glows if the circuit is set up correctly. As with Circuit 16(a), the LED could be replaced with a relay and buzzer to create a useful ice warning device.

70

Pupils could be asked to combine some of the circuits investigated, and design a complete alarm circuit (with a relay) so that a buzzer switches on if any of the following occurs:

1   A washing machine causes a flood (moisture sensor);
2   An intruder opens a window (reed switch);
3   A car enters the drive (breaking a light beam shining on an LDR);
4   The temperature outside falls below zero Celsius (thermistor).

The finished circuit might look something like the one given below.

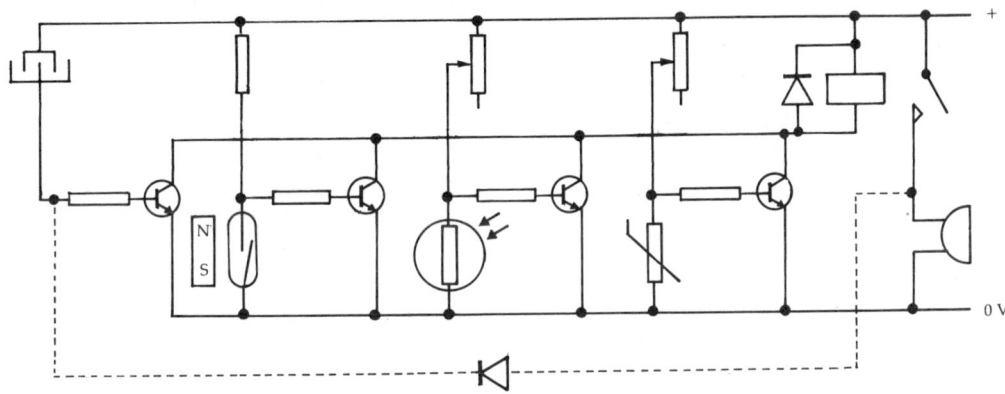

Pupils could also be asked to modify the circuit so that it latches on when triggered. The broken line in the diagram shows how a diode may be connected to achieve this. The diode allows current to flow back to the transistor base from the relay contacts after the relay has been switched on. A flying lead (without a diode) would also achieve this effect; however, current would then flow through the buzzer from the moisture sensor. This current would reduce the current flowing into the transistor base, and reduce the sensitivity of the moisture sensor.

Circuit 17(a) may be used as a 'too high' temperature warning system in, for example, a greenhouse, freezer or fridge. The circuit could be connected to a motor to open a window, or switch on a fridge or freezer motor.

Circuit 17(b) may be used as a 'too low' temperature warning system. Again this could be useful in a greenhouse, or for activating the central heating system in house. It could also detect ice in a water storage tank, or be used in a car as a road-ice alarm.

Investigation 18
Capacitor Delay

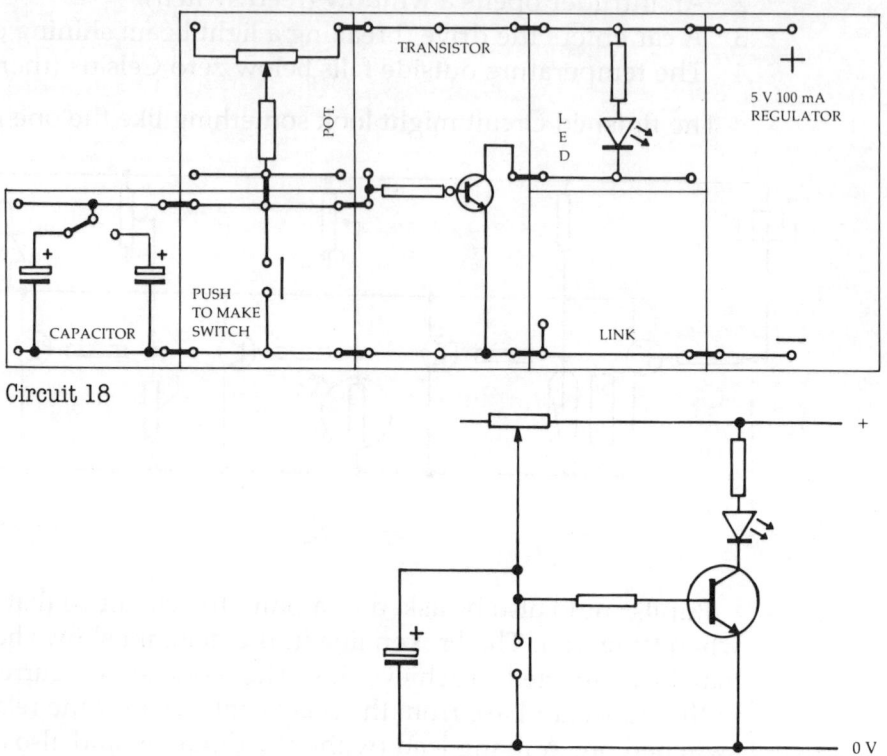

Circuit 18

Circuit 18 demonstrates how a capacitor may be employed to make a timing device, with the aid of a resistor (the pot. in this case).

When the push switch is pressed, the transistor base is connected to zero volts (or negative). The transistor and LED are therefore switched off. The capacitor is also discharged (i.e. both plates are at the same voltage – zero).

When the switch is released, current from the positive supply flows via the pot. which is wired as a variable resistor, and begins to charge the capacitor. The voltage on the top (+) plate of the capacitor rises, and when it reaches 0.6 V causes the transistor to switch on. When the pot. is turned anti-clockwise, the flow of current is reduced, causing the capacitor to take longer to charge to 0.6 V.

It should be noted that the capacitor formula $T = R \times C$ (*see* page 56) is not particularly helpful in this case, since the transistor will switch on when the voltage across the capacitor reaches 0.6 V.

Background
Information

This circuit is not an efficient timer. The low trigger voltage of 0.6 V results in the need for large capacitor values, and/or a high series resistance. Large capacitors are usually electrolytic types, with very wide tolerances (i.e. their values are not accurate). A high series resistance may not provide sufficient current to switch on the transistor properly. The problem may be reduced by using a field effect transistor (FET) which has a very high input resistance. Alternatively, CMOS logic gates (*see* page 79) provide an almost infinite input resistance, and therefore require almost no current. They also offer a much higher (two-thirds of the supply voltage) trigger voltage point.

For even longer (e.g. more than one hour) and/or more accurate timings, an 'astable multivibrator' may be employed (*see* Investigation 41). The pulses produced are fed to a counter/divider circuit which sounds the buzzer, etc., after the set number of pulses have been received. Even greater accuracy is obtained by means of an oscillating crystal (e.g. quartz). This produces a very high frequency signal, but modern counter/divider ICs are able to decode the signal and provide time displays (e.g. quartz watches).

Further Work   A high resistance voltmeter may be connected to measure the transistor base voltage (as before) whilst the capacitor charges.

The following circuit provides another variation.

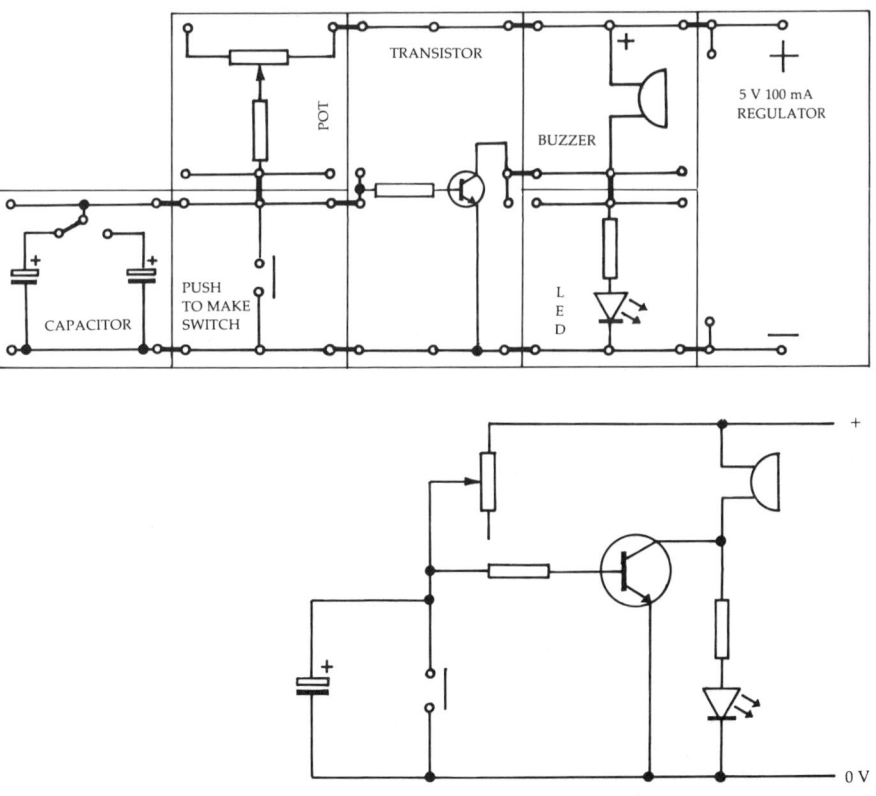

When the push switch is pressed the LED glows, but the buzzer does not sound. As the capacitor charges, the transistor begins to switch on, and the buzzer begins to sound. At the same time the LED fades out.

Pupils could be asked to explain this. They may remember Investigation 10, where a buzzer was placed in series with an LED. The buzzer will conduct enough current to allow the LED to glow, but the LED (and series resistor) will not conduct sufficiently to allow the buzzer to sound properly. However, when the transistor switches on, it allows a larger current to flow via the buzzer, but effectively 'shorts out' the LED.

Base Series Capacitor

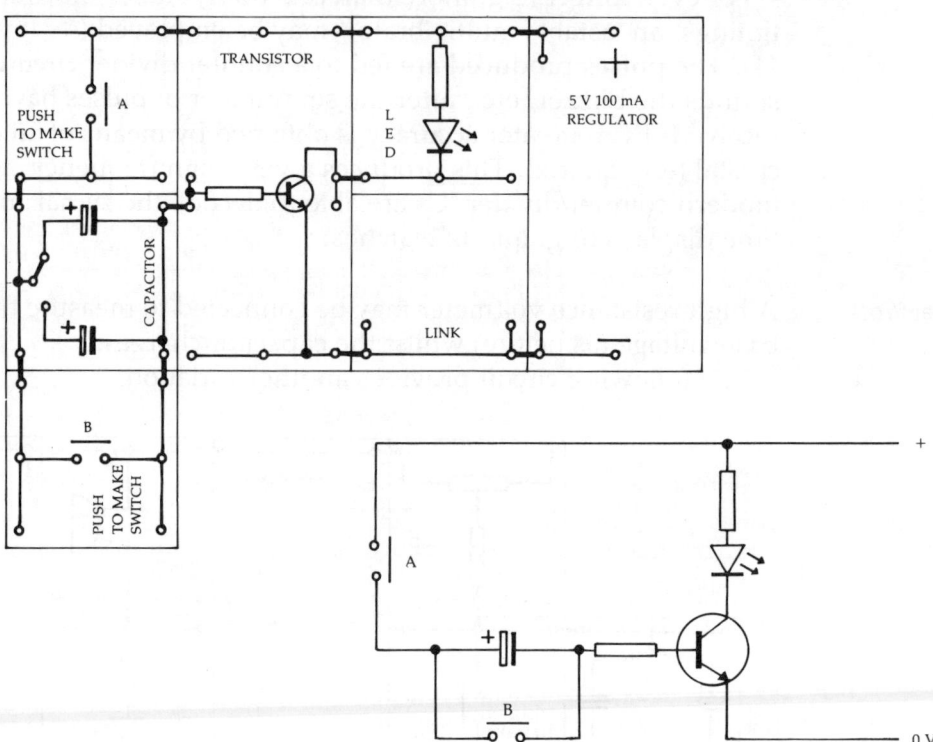

A capacitor connected in series as shown in the above diagram will provide a different effect to the one connected in parallel, and may help pupils to understand why capacitors block dc, but conduct ac.

Link the *smaller* of the two capacitors to begin with. Press switch B, to ensure the capacitor is discharged. Now press and hold down switch A. The LED should glow for a few seconds, then fade out. This occurs because current flows into the positive side of the capacitor as it charges, causing current to flow out of the negative side. This causes the transistor to switch on. When the capacitor is fully charged, no more current flows and the transistor switches off. Thus a steady dc supply is 'blocked' (after a short time), but an alternating current would continually charge and discharge the capacitor, and appear to be conducted.

The experiment may be repeated by pressing switch B to discharge the capacitor, then pressing switch A again. The larger capacitor can also be tried, but the charging time may be rather long.

Pupils could be asked to use the capacitor delay as a timing circuit to enable the alarm circuit (built in previous investigations) to be delayed after being switched on. This would allow the occupant to leave the house without setting off the alarm. Pupils could be encouraged to link the capacitor delay circuit to the alarm circuit via another relay, rather than attempt to connect transistor circuits together directly.

The following is a suitable circuit.

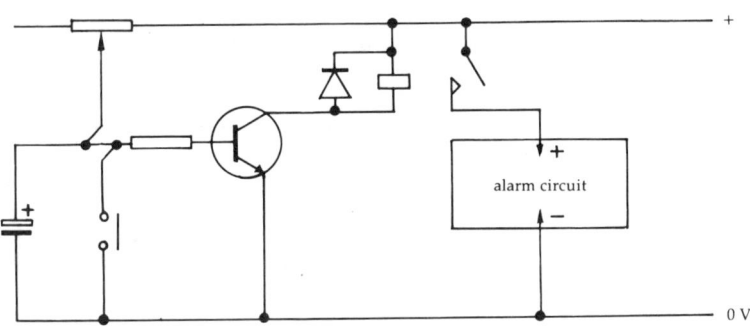

The obvious timing devices have already been mentioned. Delays are often required in electronic circuits, for example to ensure that one part of a circuit becomes active before another part, in order to provide a predictable result at switch on. Computers, printers, daisy wheel typewriters and similar devices enter a timed start-up sequence at 'power on'.

Investigation 19
Driving a Relay

When the push switch is pressed, the relay switches on (a click should be audible) and the buzzer sounds. The diagram shows the 'make' contact (sometimes called a 'normally open' contact) connected to the buzzer via a link (M). Pupils could try removing this link and using a flexible lead to link the 'break' contact ('normally closed' contact) to the buzzer. The buzzer should then sound when the push switch is not pressed, and stop sounding when the switch is pressed (see Circuit 6(b) on page 34).

This is basically a circuit which demonstrates how a relay is connected to a transistor. It could be used as a moisture detector as in Investigation 14, particularly as the buzzer works more effectively on 5 V when driven via the relay.

The use of a relay ensures that a very low resistance path is provided for the current through the relay switch contacts, and that the switch

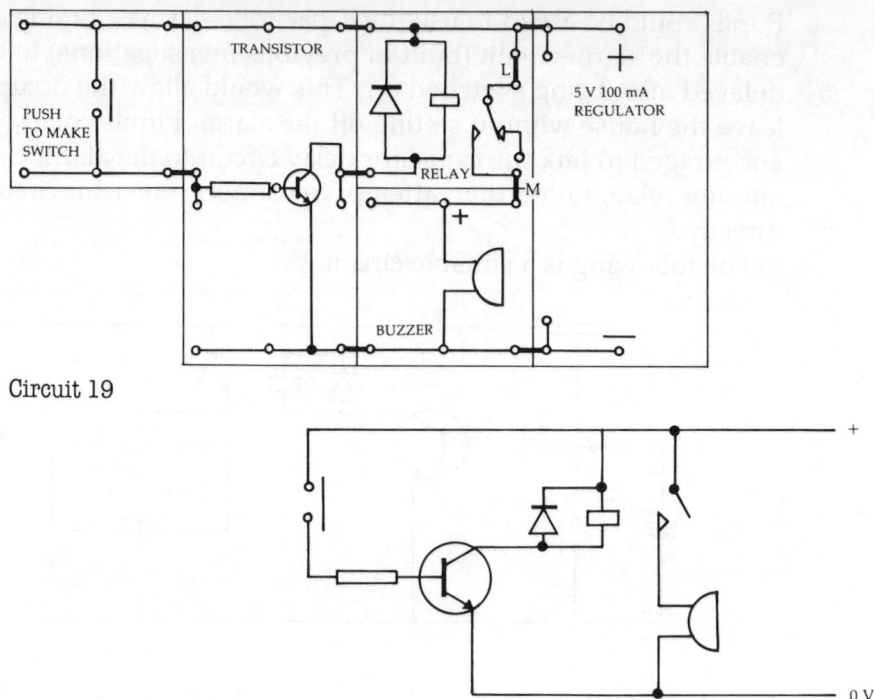

Circuit 19

contacts are electrically isolated from the relay coil. Thus, high (e.g. mains) voltages can be switched with no fear of damage to the electronic circuit or to the person using it.

It should be noted that in Circuit 19 the relay switch contacts are connected to the same supply as the rest of the circuit (via link L to positive, and via the buzzer to negative). Thus they are *not* isolated from the electronic circuit. If isolation is required, link L must be removed. Another supply (e.g. a 12 V dc supply) may then be connected, with positive connected to the relay centre contact, and negative to the negative rail of the regulator, as shown below.

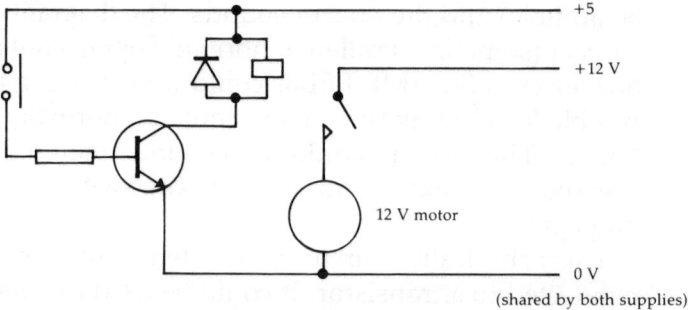

If a dangerous voltage, for example the mains supply, is to be switched, a suitable relay must be obtained and the negative supply rail must *not*

76

be shared. *However, pupils must not be allowed to experiment with mains supplies, therefore the following circuit is for information only.*

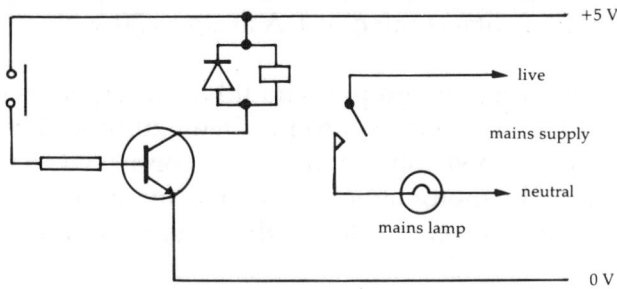

Investigation 20
Combining Circuits 16(a) and 19

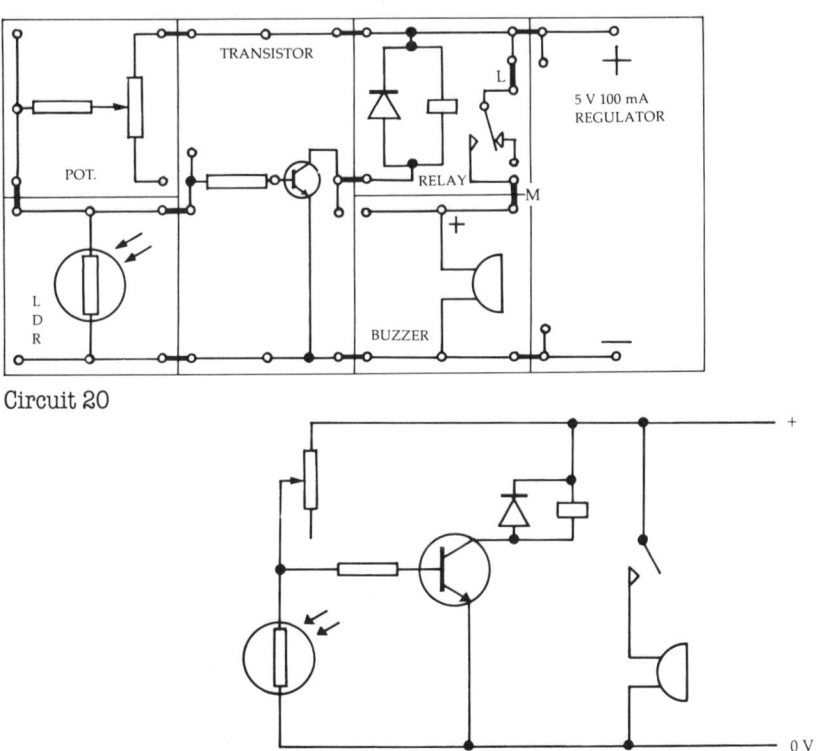

Circuit 20

In practice, the push switch in Circuit 19 will be replaced with other components. For example, any of the previous circuits may be used with the relay connected in place of the LED. The purpose of this investigation is simply to demonstrate to pupils how this might be done.

Circuit 20 shows how the light sensor transistor circuit, Circuit 16(a), could be connected to the relay. A relay allows the transistor to control any electrical device, providing that the relay contacts are capable of switching the required current and voltage. The relay supplied with the Kit has contacts rated at 1 A at up to 30 V.

Problem Solution The thermistor and pot. should be arranged as in Circuit 17(a). Links M and L on the relay board in Circuit 20 should be removed, and a separate power supply (positive) connected to the centre contact of the relay. The motor should be connected in place of the buzzer, with its negative side connected to the negative of the separate power supply.

INTRODUCTION TO LOGIC CIRCUITS

The use of digital circuitry is widespread, and encompasses computers, digital watches, calculators, robots, compact disc players and even some washing machines. Integrated circuits are capable of making decisions, where information at the inputs can be acted upon by the electronics, and a decision made via the output. Digital electronics employs two voltage levels, unlike analogue electronics, where the voltage may be at any level.

These two voltage levels are often called 'low' and 'high', where 'low' represents logic 0, and 'high' represents logic 1. In practice, logic 1 is near the positive supply voltage (often 5 V), and logic 0 is near zero volts (often the negative).

Digital circuits are made up of logic gates. These are switching circuits, the inputs and outputs of which can be represented by a truth table, as described later.

Inputs

Logic gates may have more than one input. An input should not be left 'floating' (i.e. not connected), because in some types of logic integrated circuits 'floating' inputs may change from low to high and back again very rapidly, resulting in a large consumption of current and general interference. An input must be made high or low, either by connecting it to another point in the circuit which has a clearly defined voltage and hence logic level (e.g. the output of another gate), or to the positive or negative rails (sometimes via a resistor).

In the Kent Electronics Kit, 'pull down' resistors are fitted on the NAND and NOR gate inputs to create logic 0 if an input is left unconnected. Push button switches may then be used to connect the inputs to the positive supply to create logic 1. When the push button switch is not pressed, the input is at logic 0, and when the switch is pressed the input is at logic 1. It should be stressed that the use of push button switches is not meant to imply a state of 'on' or 'off'. When the button is not pressed the 'pull down' resistor causes the input to go low. When the button is pressed, the input goes high. It is never 'off'.

The 74HC (CMOS) ICs used in this kit require an input current of less than 1 μA. The current is so low that the input resistance can be considered infinite. If the IC is removed from its socket, care must be taken not to touch the pins as you may be charged with static electricity. The high input resistance of the IC will prevent the static charge dispersing, and the high static voltage which remains may result in damage to the IC. The 'pull down' resistors fitted on the boards will prevent any damage caused by static electricity once the ICs are safely plugged into their sockets on the board.

Outputs

The logic level at an output may be determined by a voltmeter, but the 74HC (CMOS) ICs used in this kit provide sufficient current to operate an LED directly. Note that the output should not be described as being 'on' or 'off'; it is either 'high' or 'low'. It may be likened to a two-way switch, with its centre pole as the output, its 'upper pole' connected to the positive rail, and its 'lower pole' connected to the zero volts rail.

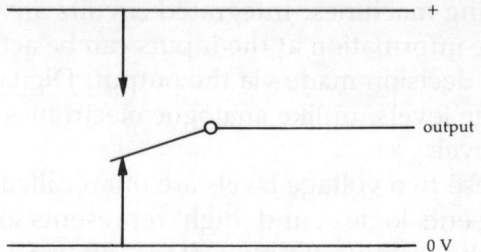

Outputs may therefore be used as a 'source' (with the LED connected between the output and zero volts rail), or as a 'sink' (with the LED connected between the positive rail and the output). When used as a source the LED will glow when the output is at logic 1. When used as a sink the LED will glow when the output is at logic 0. Note that an LED should always be used with a series resistor; this is already fitted to the LED boards in the Kent Electronics Kit.

Completing Truth Tables

The purpose of a truth table is to show all the possible combinations of logic levels (i.e. logic 0 and logic 1) at the inputs and outputs. Investigations 21 to 26 require the results to be expressed as a truth table, with the output wired as a source.

The convention used is as follows:

Input push button not pressed   — input at logic 0
Input push button pressed and held down — input at logic 1
Output LED not glowing   — output at logic 0
Output LED glowing   — output at logic 1

Using a Logic Gate Board

The circuits described in the investigations which follow represent a few of the possibilities available. There is great scope for combining circuits or making small changes to create a different effect. Pupils are often very good at this kind of experimentation. For example, having wired up Circuit 29(a), many pupils should be able to devise Circuit 29(b) when asked to make the circuit behave in the opposite way.

Pupils may be allowed to experiment, with a very low risk of damage to the kit, providing that flying leads are not used to bypass the regulator. Always ensure that the power rails (both positive and negative) are connected properly, right up to the logic gates. Removing the positive or negative supplies to the gates is unwise, and will produce unpredictable results.

Investigation 21
Truth Table for a NAND Gate

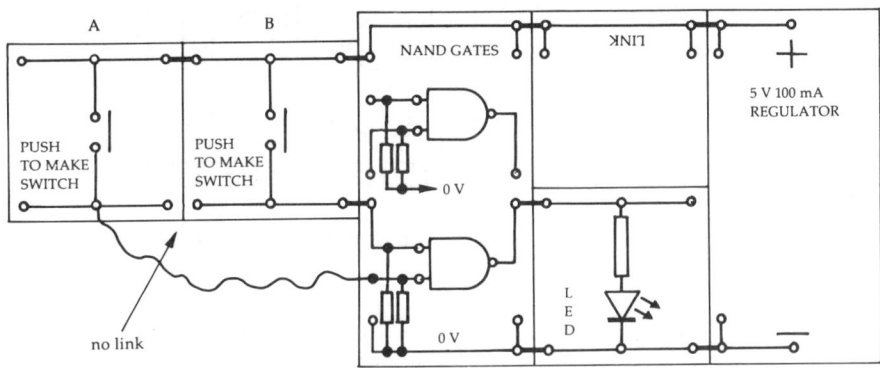

Circuit 21

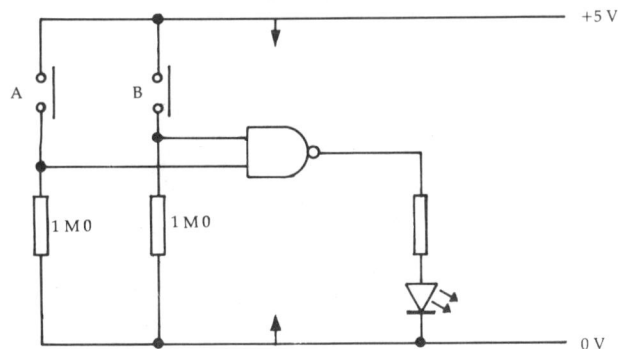

$$+5\,V$$

$$0\,V$$

Note: In this and the following circuits, the **1 MΩ** resistors are the ones already fitted to the NAND gate and NOR gate boards.
The arrows [ ] indicate that the gates are also internally connected to the supply rails.

The lower NAND gate is used in this circuit, the two inputs being controlled by means of push to make switches.

When wired correctly, if neither switch is pressed, the LED should glow. The LED should continue to glow if either push switch is pressed, but if *both* switches are pressed at the same time the LED should stop glowing.

The truth table obtained should be as follows:

Input A	Input B	Output
0	0	1
1	0	1
0	1	1
1	1	0

81

Background Information	The IC used in the Kent Electronics Kit is a type 74HC00 and includes four NAND gates, each with 2 inputs.

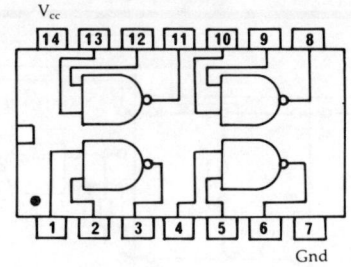

74HC00 Quadruple 2-input NAND gate

NAND gates are available with more than two inputs, such as type 74HC10 with three 3-input NAND gates, 74HC20 with two 4-input NAND gates, 74HC30 with one 8-input NAND gate, and 74HC133 with one 13-input NAND gate.

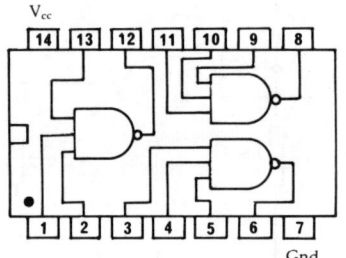

74HC10 Triple 3-input NAND gate

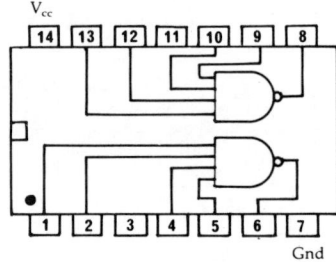

74HC20 Dual 4-input NAND gate

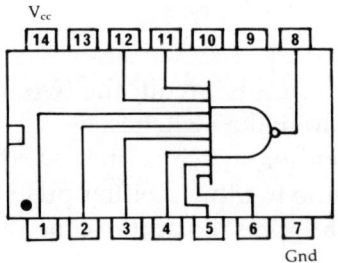

74HC30 8-input NAND gate

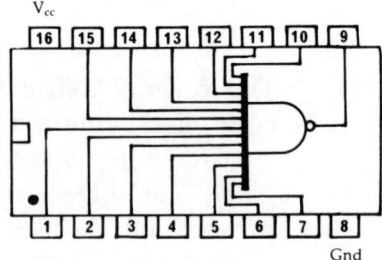

74HC133 13-input NAND gate

Further Work	It is possible to create a NAND gate using a combination of NOR gates. After completing the next investigation, pupils could be asked to design a NAND gate using NOR gates. This could be tested by borrowing an additional NOR board. The solution is given on the next page.

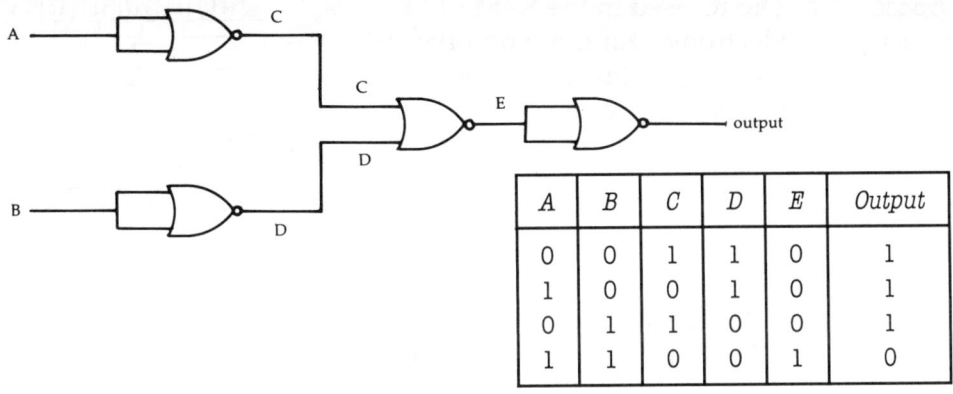

A	B	C	D	E	Output
0	0	1	1	0	1
1	0	0	1	0	1
0	1	1	0	0	1
1	1	0	0	1	0

Investigation 22
Truth Table for a NOR Gate

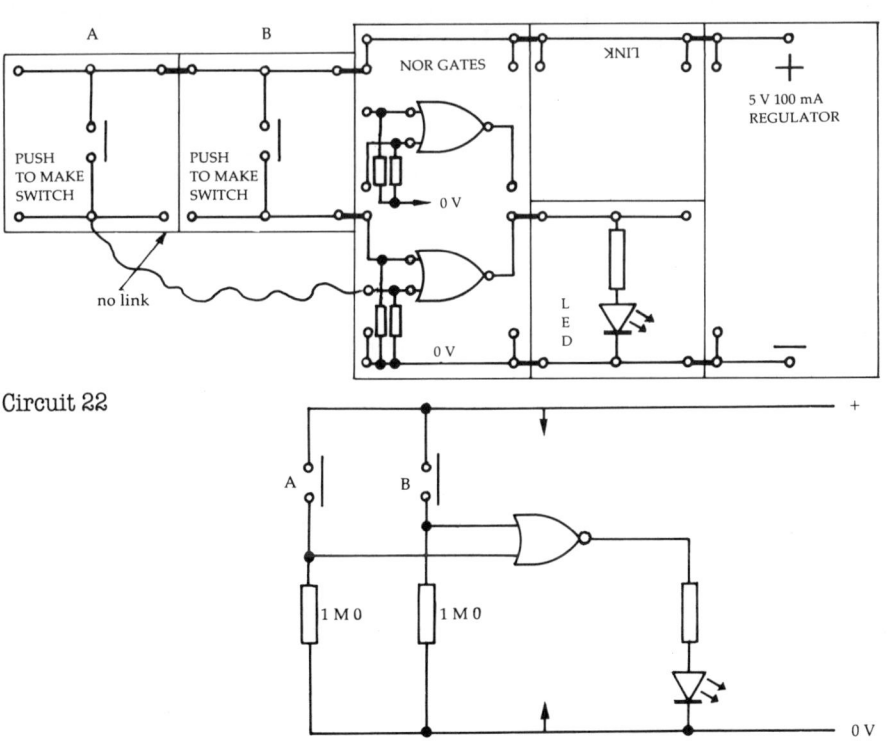

Circuit 22

The output LED should normally glow. When either or both switches are pressed the LED should stop glowing. The truth table on the right should be obtained.

Input A	Input B	Output
0	0	1
1	0	0
0	1	0
1	1	0

Background Information	The IC used in the Kent Electronics Kit is a type 74HC02 which contains four NOR gates in a single package.	

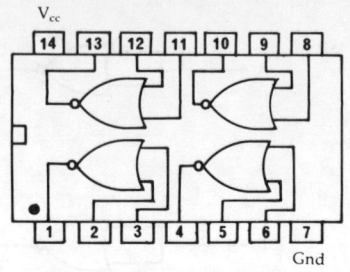

74HC02 Quadruple 2-input NOR gate

NOR gates may be obtained with more than two inputs, such as the 74HC27 IC with three 3-input NOR gates, the 74HC4002 IC with two 4-input NOR gates, or the 74HC4078 IC with one 8-input NOR gate.

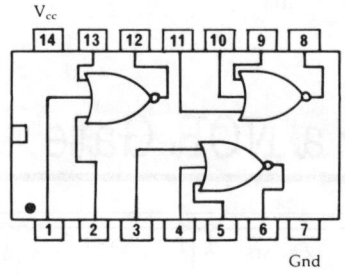

74HC27 Triple 3-input NOR gate

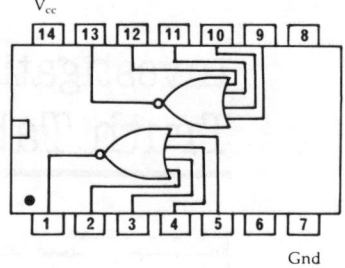

74HC4002 Dual 4-input NOR gate

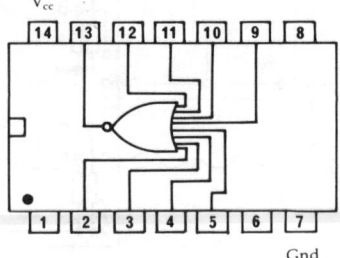

74HC4078 8-input NOR gate

Further Work/Homework	A NOR gate may be produced using a combination of NAND gates. Pupils could be asked to design a suitable circuit, which could then be tested by borrowing an additional NAND board. The solution is given below.

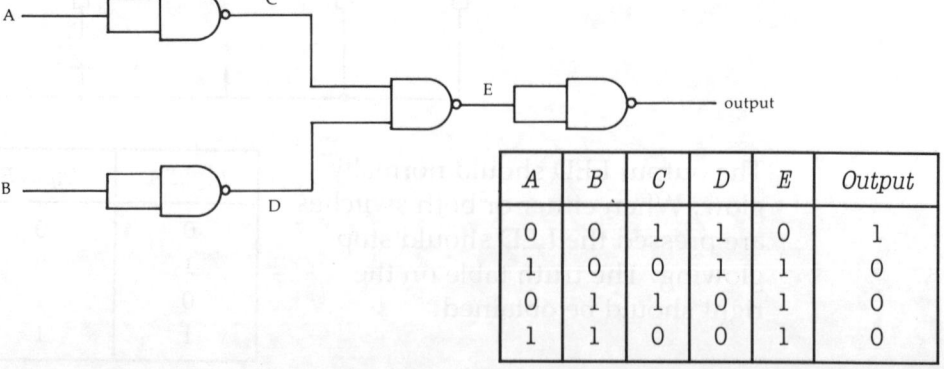

A	B	C	D	E	Output
0	0	1	1	0	1
1	0	0	1	1	0
0	1	1	0	1	0
1	1	0	0	1	0

84

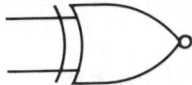

An exclusive NOR gate, shown above (not included in the Kit), causes the output to go high only if the two inputs are at the same state. If the two inputs are different, the output goes low. The truth table is as shown opposite.

. Input A	Input B	Output
0	0	1
1	0	0
0	1	0
1	1	1

The IC type 74HC266 includes four 2-input exclusive NOR gates in one package.

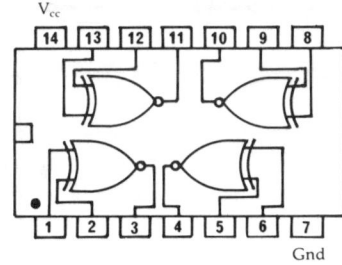

74HC266 Quadruple 2-input exclusive NOR gate

Further Work/Homework

An exclusive NOR gate may be produced by using a combination of NAND gates. Pupils could be asked to design a suitable circuit, which could be tested by borrowing additional NAND gate boards.

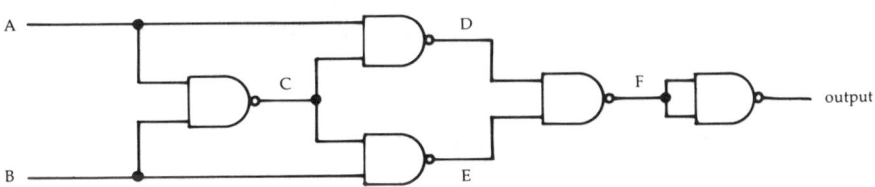

A	B	C	D	E	F	Output
0	0	1	1	1	0	1
1	0	1	0	1	1	0
0	1	1	1	0	1	0
1	1	0	1	1	0	1

Investigation 23
Truth Table for a NOT Gate

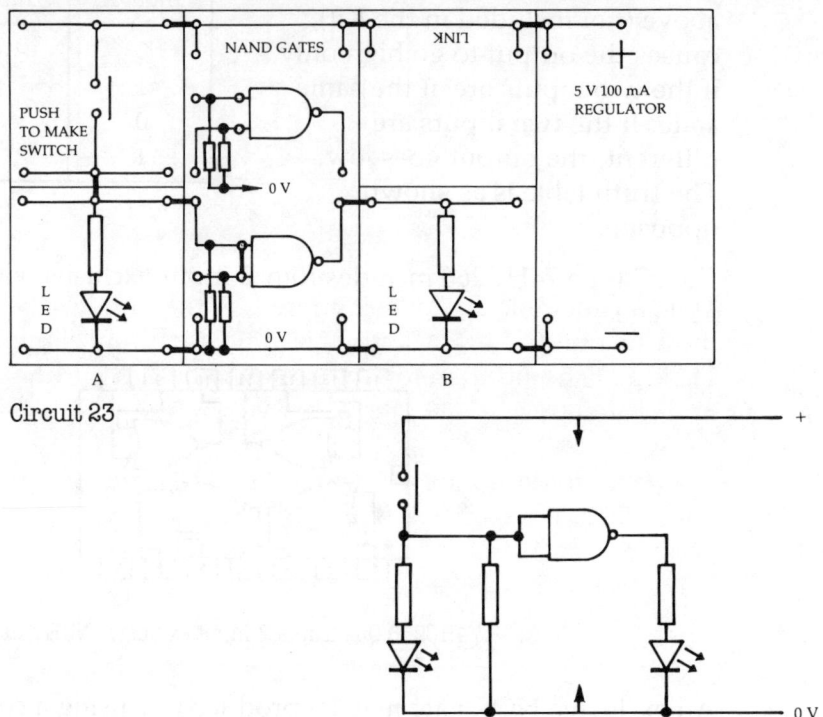

Circuit 23

The NAND gates board is used for this investigation. However, the NOR gates board may be substituted, if desired.

When the push switch is not pressed, LED A should be off, and LED B on. When the switch is pressed, LED A should glow, and LED B should switch off. Thus the circuit demonstrates that the output from a NOT gate is at the opposite logic level to the input.

The truth table given opposite should be obtained.

Input A	Output B
0	1
1	0

The NOT gate (often called an inverter) has been produced by connecting the NAND gate inputs together. The lower NAND gate is employed in this circuit, and LED A is used to indicate when the push switch is pressed.

In practice, a NOT gate IC could be used instead of linking a pair of NAND gate inputs together. A NOT gate is drawn as shown opposite.

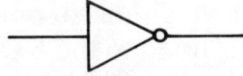

NOT gates can be purchased as integrated circuits, such as IC type 74HC04 which includes six NOT gates (inverters) in one package.

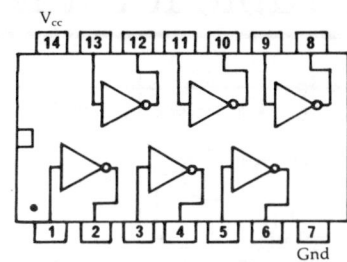

74HC04 Hex inverter

Further Experiments

The considerable advantage offered by the high resistance input (*see* Introduction to Logic Circuits) is almost entirely eliminated by LED A. This LED should therefore be removed before further experiments are undertaken.

Further Work/Homework

If the link joining the two inputs together is removed, the circuit no longer behaves like a NOT gate. Pupils could be asked to remove this link, then add a flying lead (in a different place) to make the NAND gate behave like a NOT gate. The solution is given below.

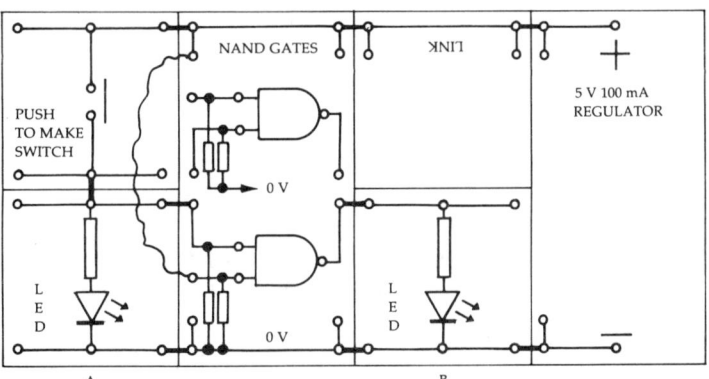

The NOR gates board may be substituted in this investigation, since joining the NOR gate inputs together also creates a NOT gate. Pupils could be asked to try out the NOR gates board and find out if the results are the same.

Having established this, pupils could then remove the link connecting the two inputs together. The NOR gate will still behave like a NOT gate. Some pupils may be able to explain this. The reason is that the unused input is held at zero volts (logic 0) by one of the resistors fitted to the NOR gates board. A glance at the NOR gate truth table will reveal that if one input is at logic 0, the output will be at the opposite logic level to the other input.

Investigation 24
Truth Table for a Buffer Gate

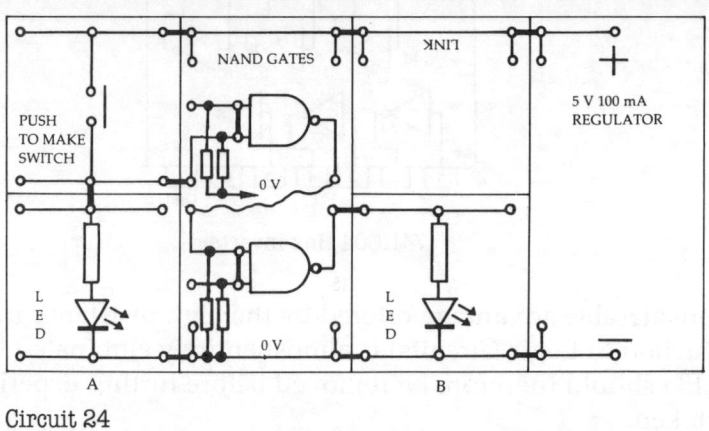

Circuit 24

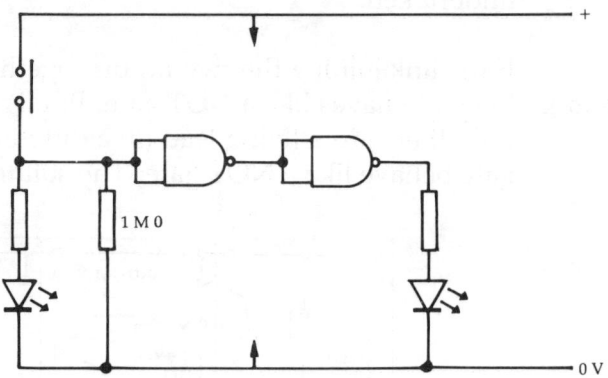

Note: The 1 MΩ resistor represents the ones already fitted in the NAND gates board.
The similar pair of resistors on the second gate are superfluous and therefore not shown.

The NAND gates board is used for this investigation. However, the NOR gates board may be substituted, if desired.

When the push switch is not pressed, neither LED should glow. When the push switch is pressed, both LEDs should glow.

The truth table given opposite should be obtained.

Input	Output
0	0
1	1

Investigation 23 showed how the inputs of a NAND gate may be connected together to make a NOT gate. If the output from one NOT gate is connected to the input of a second NOT gate as shown below, the logic level of the final output will mirror the logic state of the original input.

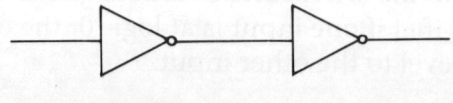

This type of circuit is known as a 'buffer' gate, and is normally drawn as shown opposite.

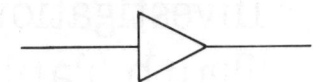

Background Information

At first sight the buffer may not appear to be particularly useful, but it should be noted that the current required at the input to the buffer is very small (less than 1 μA), yet this is able to control a current of many milliamps from the output. The buffer therefore acts like a current amplifier, except that this type of circuit is designed to work with two distinct voltage levels rather than a continuously varying voltage as in, for example, an audio amplifier.

Buffer gates may be purchased in integrated circuit form, such as IC type 74HC4050, which contains six buffer gates in one package.

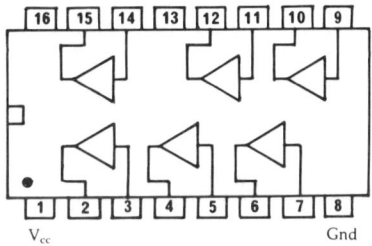

74HC4050 Hex buffer

As with the NOT gate in the previous investigation, the LED connected below the push switch eliminates the advantage offered by the gate's very high input resistance. Thus, before attempting further experiments, the input LED should be removed from the circuit.

Further Work/Homework

The NOR board may be substituted to allow pupils to find out if the results are identical. As with the previous investigation, the links connecting the inputs together on the NAND gates or the NOR gates could be removed, and pupils asked to find an alternative way of making the gates behave as buffers (*see* Investigation 23 'Further Work/Homework').

Since a NAND or NOR gate can behave as a NOT gate, a NAND gate could be followed by a NOR (and vice-versa) to create a buffer.

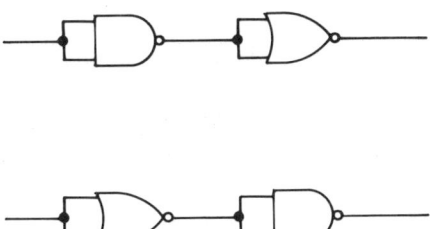

Both logic boards could be connected as buffers, and pupils asked to find out if there is any advantage in using two buffers in series. There is little advantage gained, and the circuit will work as with one buffer.

Investigation 25
Truth Table for an AND Gate

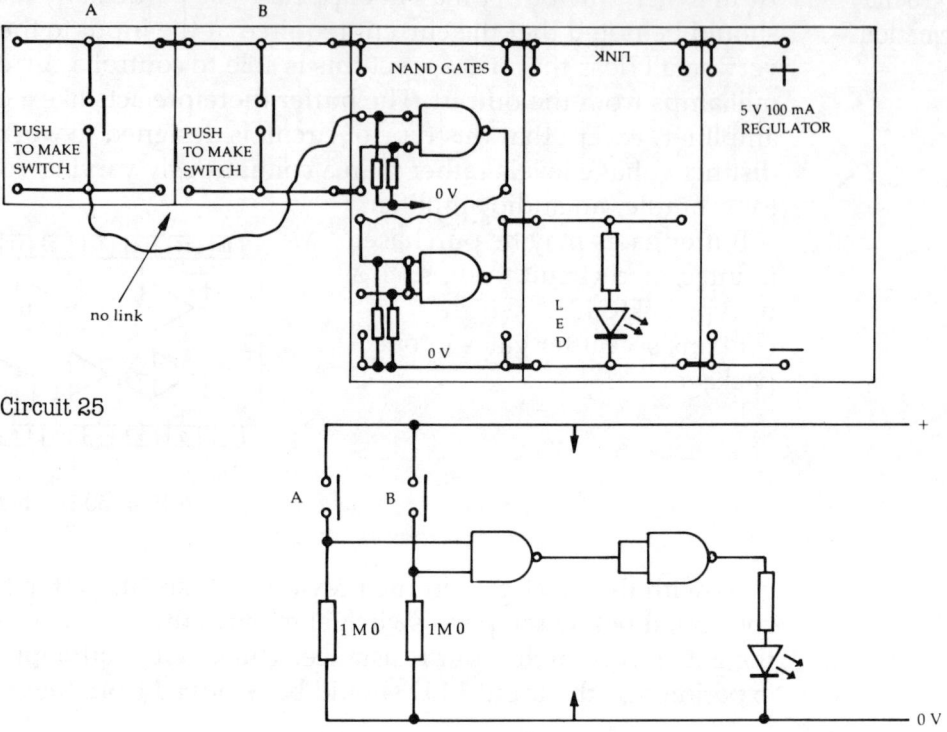

Circuit 25

Note: The resistors on the second gate are superfluous and therefore not shown.

In Circuit 25 the output LED should glow only when *both* push switches are pressed. The truth table given below should be obtained.

Input A	Input B	Output
0	0	0
1	0	0
0	1	0
1	1	1

In this circuit, the NAND gate is followed by an inverter (NOT gate), produced by connecting the lower NAND gate's inputs together.

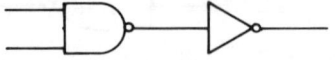

Thus the action of the NAND gate is inverted, and an AND gate is produced.

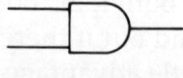

90

Background
Information

AND gates may be purchased in integrated circuit form, such as IC type 74HC08 which includes four AND gates in one package.

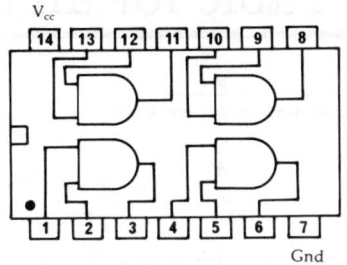

74HC08 Quadruple 2-input AND gate

Further
Work/Homework

An AND gate may be produced by a combination of both NAND and NOR gates, or by means of several NOR gates. Pupils could design suitable circuits and test them out. Some possible solutions are given below.

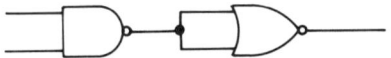

NAND followed by NOR connected as an inverter

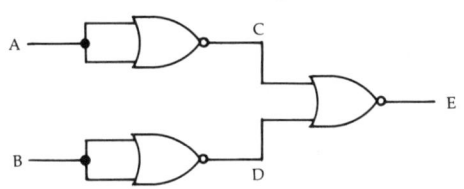

A	B	C	D	E
0	0	1	1	0
1	0	0	1	0
0	1	1	0	0
1	1	0	0	1

Investigation 26
Truth Table for an OR Gate

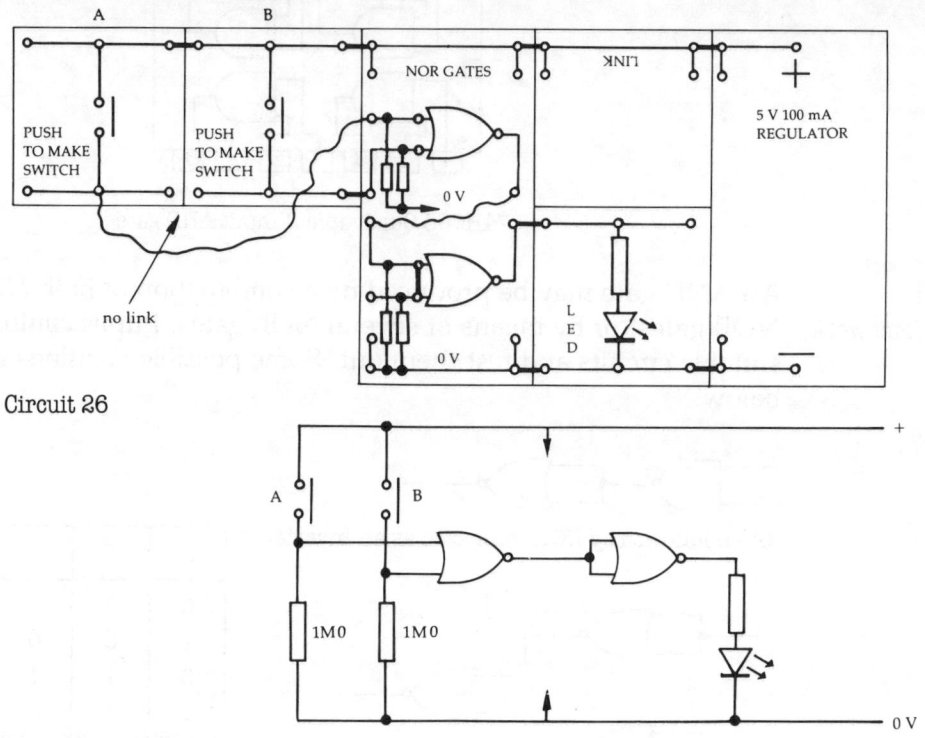

Circuit 26

The circuit demonstrates that if neither switch is pressed, the output LED remains off. If one switch, *or* the other, *or* both are pressed, the LED glows.

The truth table given on the right should be obtained.

Input A	Input B	Output
0	0	0
1	0	1
0	1	1
1	1	1

In this circuit, the lower NOR gate inputs are connected together to make a NOT gate (inverter) in the same way as the two inputs to the NAND gate are joined together in Investigation 23. When this inverter is connected to the output of a NOR gate, the effect of the NOR is reversed, making an OR gate.

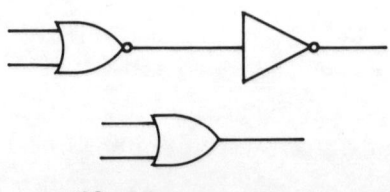

92

Integrated circuit OR gates are available, for example IC type 74HC32 which includes four OR gates in one package.

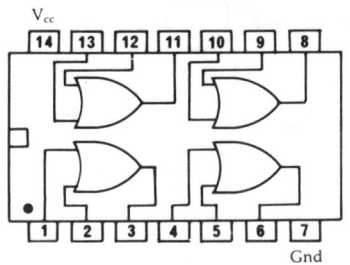

74HC32 Quadruple 2-input OR gate

The OR function may be accomplished with diodes. Almost any number of inputs are possible, simply by adding more diodes. Connecting the diode cathodes to a buffer gate provides a perfectly usable 'OR gate'.

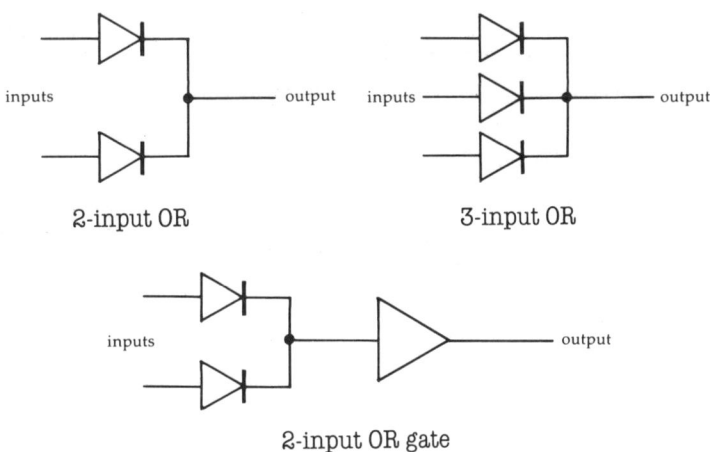

2-input OR   3-input OR

2-input OR gate

OR gates are available with more than two inputs. For example, IC type 74HC4075 contains three 3-input OR gates in one package.

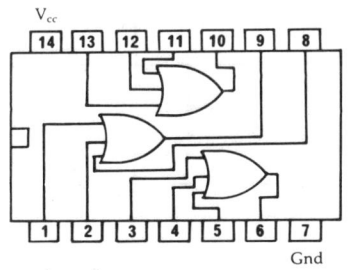

74HC4075 Triple 3-input OR gate

An OR gate can be created by using a combination of NOR and NAND gates, or by combining three NAND gates. Pupils could be asked to design and test suitable circuits. Some possible solutions are given below.

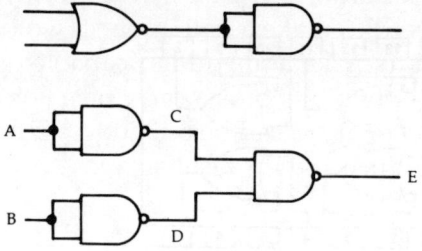

A	B	C	D	E
0	0	1	1	0
1	0	0	1	1
0	1	1	0	1
1	1	0	0	1

Exclusive OR Gate

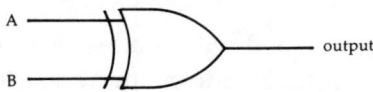

The exclusive OR gate, shown above (not included in the Kit), causes the output to go high if one *or* other input is high. However, if *both* inputs are high the output goes low. The truth table is shown on the right.

Input A	Input B	Output
0	0	0
1	0	1
0	1	1
1	1	0

The IC type 74HC86 includes four 2-input exclusive OR gates in one package.

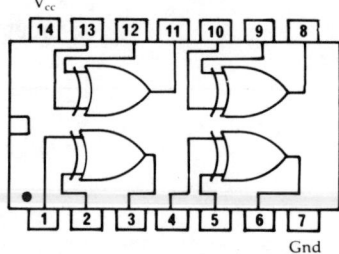

74HC86 Quadruple 2-input exclusive OR gate

Pupils could be asked to design an exclusive OR gate by combining four NAND gates. The solution is given below.

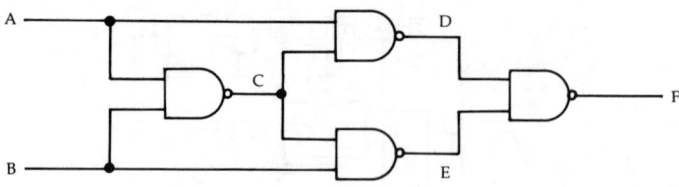

A	B	C	D	E	F
0	0	1	1	1	0
1	0	1	0	1	1
0	1	1	1	0	1
1	1	0	1	1	0

Exclusive OR gates are useful as 'difference circuits', since they produce a high output only if the inputs are at different states. Their chief application is in binary addition circuits, where an exclusive OR gate and an AND gate are used to make a half-adder. Two half-adders and an OR gate are required to make a full-adder, capable of adding together three binary bits (i.e. two bits plus a carry bit). For example a 4-bit binary number 1110 (decimal 14) can be added to a 2-bit binary number 11 (decimal 3) to produce 10001 (decimal 17).

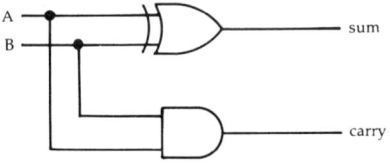

Half-adder circuit

A	B	Sum	Carry
0	0	0	0
1	0	1	0
0	1	1	0
1	1	0	1

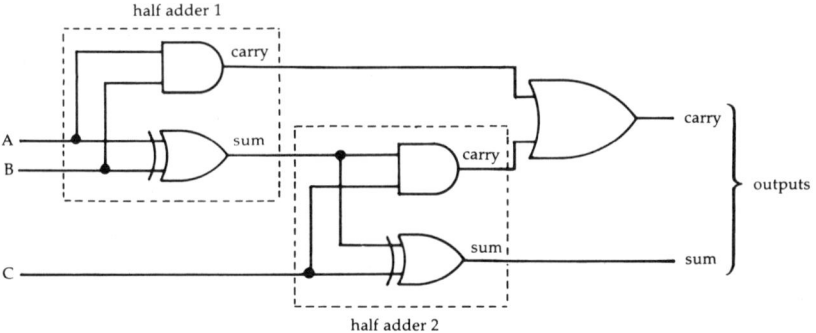

Full-adder circuit

| Inputs | | | Outputs | |
A	B	C	Sum	Carry
0	0	0	0	0
0	0	1	1	0
0	1	0	1	0
1	0	0	1	0
0	1	1	0	1
1	0	1	0	1
1	1	0	0	1
1	1	1	1	1

Investigation 27
Using the Output as a Sink

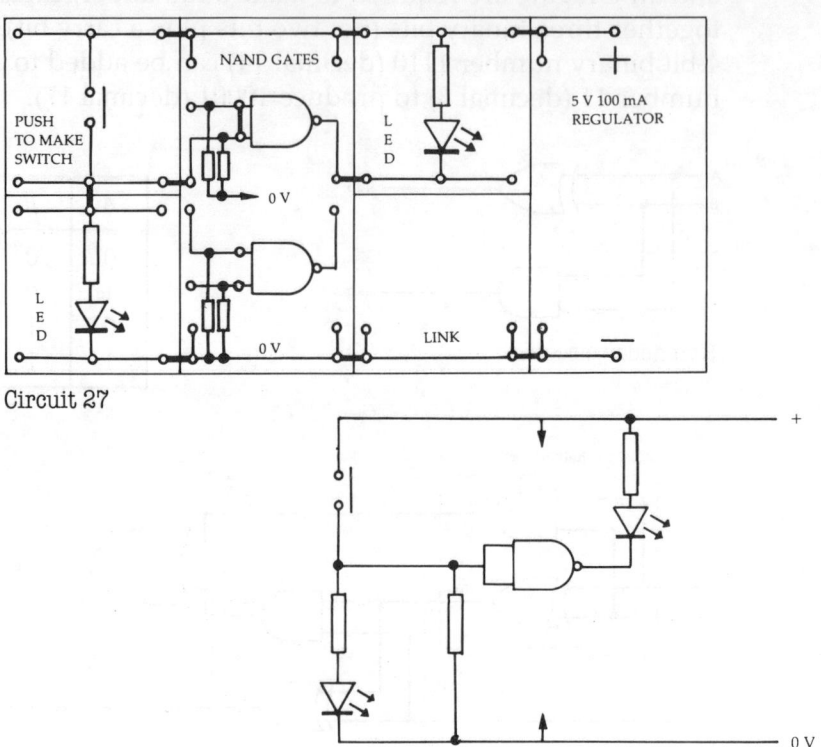

Circuit 27

In Circuit 27, when the push switch is off, both LEDs should be unlit. When the push switch is pressed, both LEDs should glow.

The circuit demonstrates that outputs of the 74HC series ICs act equally well as a current 'source' or a current 'sink'. The output LED lights when the output is *low*. When the output is high the output LED stops glowing. Great care is needed in practice with this type of circuit, as it causes much confusion (e.g. LED glowing indicates logic 0). The circuit further illustrates that the output is never 'on' or 'off', but is 'high' or 'low', and can light a suitably wired LED in either state.

Applications

It is sometimes more convenient to drive an LED in this way, and in some circuits this method would save having to use another gate to invert the output. Using the outputs as current sinks was popular when the original TTL 74 series ICs became available. Both these and the later 74LS series were able to sink a much higher current than they could source. The later 74HC ICs used in the Kent Electronics Kit are able to source and sink equal amounts of current.

When the output is used to switch a transistor, a logic 1 output

96

switches on an *n-p-n* transistor (e.g. type BC184L as used in the kit). The opposite can be achieved by using a *p-n-p* transistor (such as the type BC214L). A logic 0 will switch on this type of transistor.

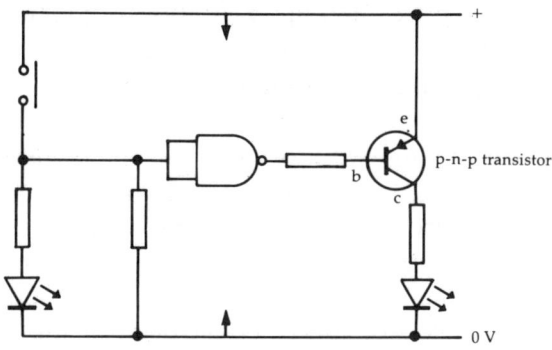

Investigation 28
Moisture Detector/Touch Switch

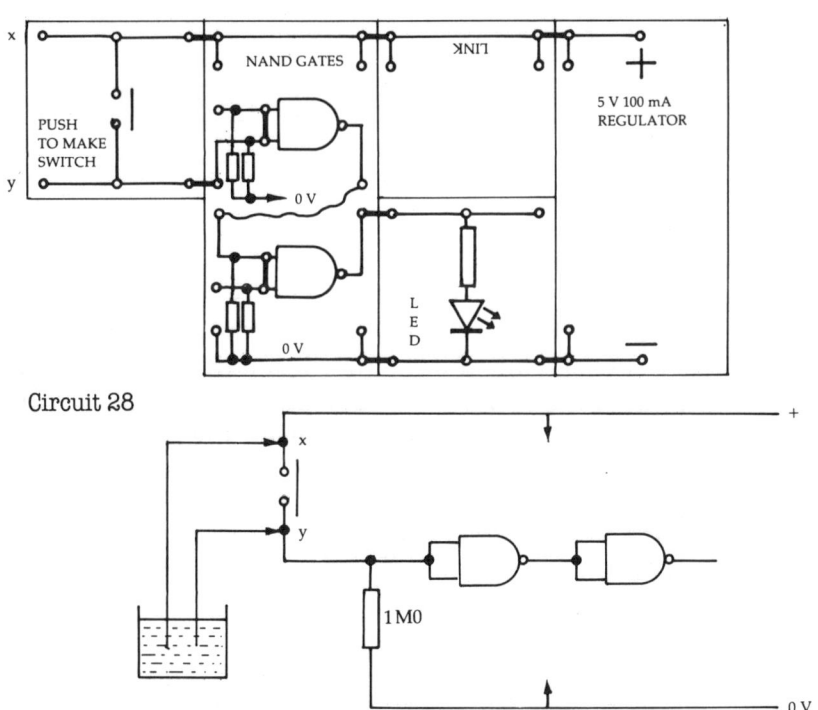

Circuit 28

Note: The resistors on the second gate are superfluous and therefore not shown.

Circuit 28 is very similar to the buffer circuit already described in Investigation 24, except that the LED below the push switch is omitted. This is important, as the moisture detector and touch switch

97

applications require a very high input resistance at the buffer. In other words, very little current is available from the moisture sensor or touch switch contacts, and an LED connected across the buffer inputs would bypass this current to the zero volts rail.

The push switch illustrated should be regarded as a test switch, and a pair of flying leads would be connected as shown to connect with the moisture sensor or touch contacts.

Moisture Sensor A pair of bare wires placed over the edge of a container will detect the presence of many liquids, such as water, by conducting a small electric current through the liquid. This current, though small, is sufficient to cause the voltage at the buffer's input to rise to logic 1. It is important to ensure that the two bare wires do not touch each other.

A rain sensor can be constructed using strips of foil, or a small piece of stripboard, as shown on page 61.

Touch Contacts A pair of touch contacts can be made from any pieces of metal, such as drawing pins. These should be placed close together, so that your finger can touch both at the same time.

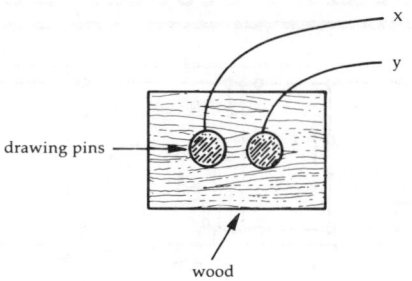

Further Work Pupils could experiment to find out how the circuit may be made to latch (i.e. stay on after being triggered). A piece of wire joining the output of the lower gate, back to the inputs of the upper gate, may appear to be the solution. However, the sensitivity of the circuit will be severely reduced, as current will then flow from the moisture or touch sensors directly to the output of the lower gate.

At this stage, some pupils should be able to suggest that a diode be used to allow the feedback action, without current flowing towards the output as described.

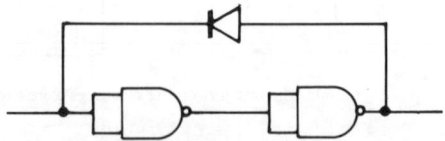

Alternatively, the NOR gates board could be substituted. The two inputs to the upper gate should *not* be linked together. The output of the lower gate may then be fed back to the spare input, as shown below.

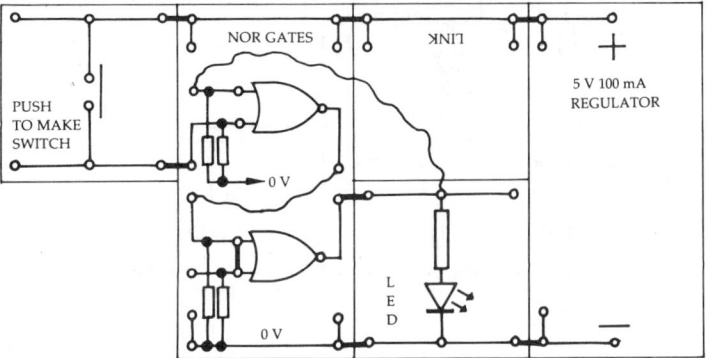

Homework Suggestion

Pupils could be asked to design a very sensitive water alarm, with or without a latch, which sounds a buzzer.

Investigation 29
Light Sensor

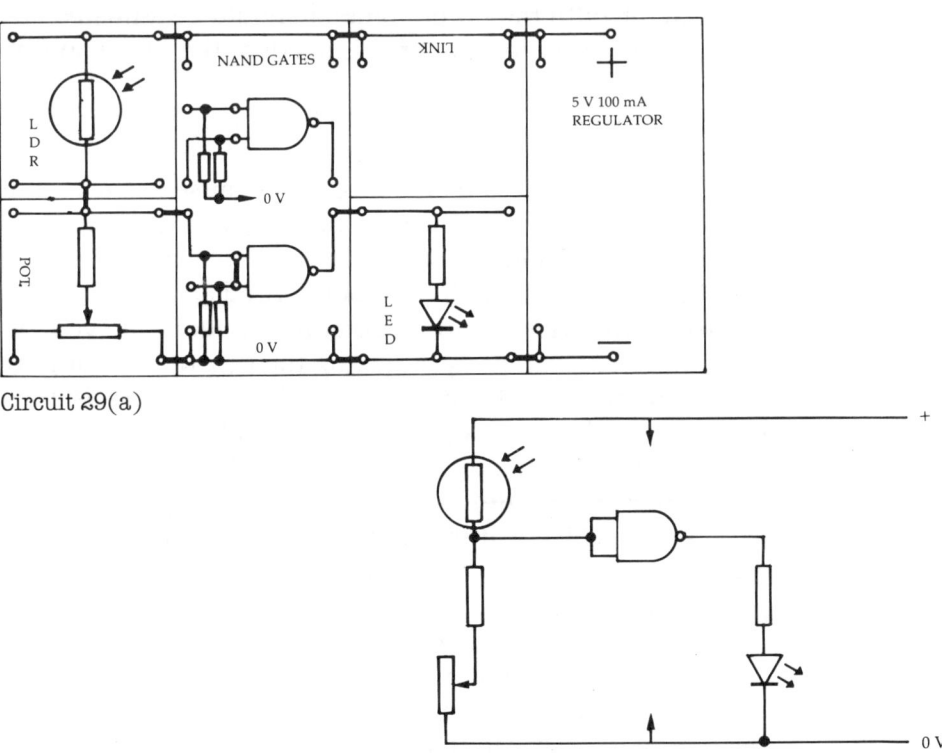

Circuit 29(a)

In Circuit 29(a) the potentiometer should be adjusted so that the LED is not quite glowing. When the LDR is shaded, the LED should glow.

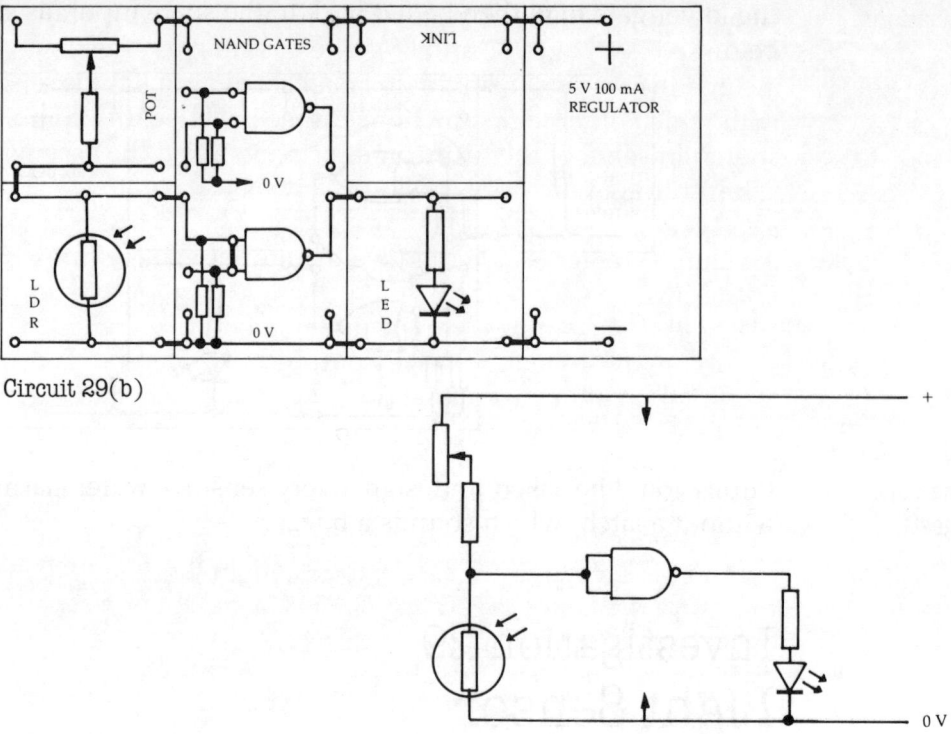

Circuit 29(b)

In Circuit 29(b) the potentiometer should be adjusted so that the LED is just glowing. When the LDR is shaded, the LED should stop glowing.

In both cases, the circuit should be sensitive to small changes in the level of light. However, when used in a classroom the sensitivity may appear somewhat less than expected, because of the amount of light reaching the LDR from a variety of directions. Switching off some, or all, of the artificial lights may produce a substantial improvement.

In Circuit 29(a) the LDR and pot. (wired as a variable resistor) are connected in series to form a potential divider. The voltage at their junction can be varied by adjusting the pot. or by changing the light falling on the LDR. As the level of light falls, the resistance of the LDR increases, and the voltage at the junction therefore decreases. The junction is connected to the input of the NAND gate, which is wired as an inverter (NOT gate).

When the voltage at the junction is low (less than about 2 V), the inverter's output goes high, lighting the LED. When the voltage at the junction is high (more than about 3 V), the inverter's output goes low and the LED stops glowing.

In Circuit 29(b) the pot. and LDR are interchanged. This causes the voltage at their junction to rise, as the light on the LDR falls. Otherwise the theory is similar to that of Circuit 29(a).

There are other ways of making the circuit behave in the opposite way. For example, the output could be used as a sink rather than a source.

General applications of light sensitive circuits have already been discussed (*see* pages 52 and 68). The advantages of using logic gates include increased sensitivity to small changes in light levels, the ease with which other logic functions can be included and combined (as shown in some of the forthcoming circuits), and the ease with which a Schmitt trigger (described below) can be incorporated.

A Schmitt trigger is able to move the output voltage from low to high, or the reverse very quickly, and provide the property known as *hysteresis*. Its usefulness can be understood by reference to the light sensor circuit. Imagine that the output LED is a room lamp, designed to switch on automatically at dusk, Circuit 29(a). As the daylight fades, a point is reached where the lamp is about to switch on. If a cloud happened to pass at that moment, the temporary reduction in daylight would make the lamp switch on. However, when the cloud had passed, the lamp may well switch off again. In this way, the lamp may switch on and off several times before the daylight level had fallen sufficiently to ensure the lamp stayed on. The problem would be even worse if the circuit was controlling a pair of curtains!

The solution is to introduce hysteresis. This means that once the circuit has switched on the lamp, the daylight level would have to rise by a substantial amount before the lamp would switch off again. Hysteresis is provided by connecting a relay to the output (via a transistor). The nature of the relay ensures that, once switched on, the current flowing through its coil must fall significantly before it switches off again. In this case the amount of hysteresis provided cannot be controlled by the circuit designer.

A NAND gate Schmitt trigger provides hysteresis, and is connected as shown below.

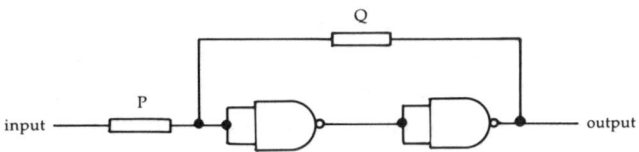

The resistor Q introduces *positive feedback*, where the voltage at the output is fed back to the input to help maintain its current state. Resistor P assists the effect by reducing the influence of the input circuitry (e.g. the LDR).

The ratio of resistors P and Q determines the amount of hysteresis provided. More advanced pupils could introduce resistor Q via crocodile leads. A value of 47 kΩ would illustrate the effect, assuming that resistor P was the 10 kΩ resistor fitted to the resistor board.

Note that both NAND gates are now in use, and are wired to form a buffer. The circuit will therefore work the opposite way round, and the Schmitt action will ensure that a larger change of light level is required to make the circuit switch the LED on or off. In other words, the circuit has become less sensitive.

If an oscilloscope is used to monitor the output voltage, it will reveal a sudden change from low to high, or high to low. If resistor Q is removed, the Schmitt action will no longer apply, and the output voltage may show a slower transistion between low and high.

Schmitt triggers are available as fully integrated circuits, for example 74HC14 (which includes six Schmitt inverter gates), or 74HC132 (which includes four 2-input Schmitt NAND gates).

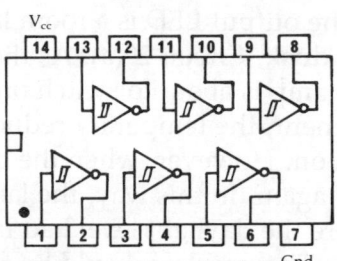

74HC14 Hex Schmitt trigger

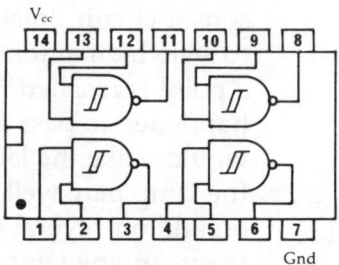

74HC132 Quadruple 2-input NAND Schmitt gate

Further Work   Pupils could be asked to add a transistor and relay to the circuit. A 6 V 60 mA lamp could then be controlled by the relay via flying leads. This could be used to simulate the addition of a mains lamp, for example a room light or street lamp.

An interesting effect may be observed if this lamp is placed close to the LDR when connected as shown in Circuit 29(a). Light from the lamp will enter the LDR which will then cause the lamp to switch off. The lack of light will effect the LDR which then causes the lamp to switch on again. The time taken for the lamp filament to heat up will introduce a delay and the lamp will therefore switch on and off very quickly. The flashing rate may be too fast to observe, but the relay will produce a series of clicks as it switches on and off. This type of feedback is common in electronics and is often unwanted. Another example of feedback sometimes occurs when a microphone is placed near a loudspeaker, resulting in an unpleasant whistle.

Homework Suggestion   In this investigation the lower NAND gate is used as an inverter, or NOT gate. Pupils could be asked to design a circuit using both NAND gates (similar to the Schmitt trigger arrangement described above) with their two inputs joined together making them NOT gates. When these NOT gates are connected in series a buffer is produced, as in Investigation 24. Using this as a starting point, pupils could design a circuit using the buffer, which would have the same effect as Circuit 29(a).

The LDR and pot. should be arranged as in Circuit 29(b), since the buffer will have the opposite effect to a NOT gate.

Investigation 30
Temperature Sensor

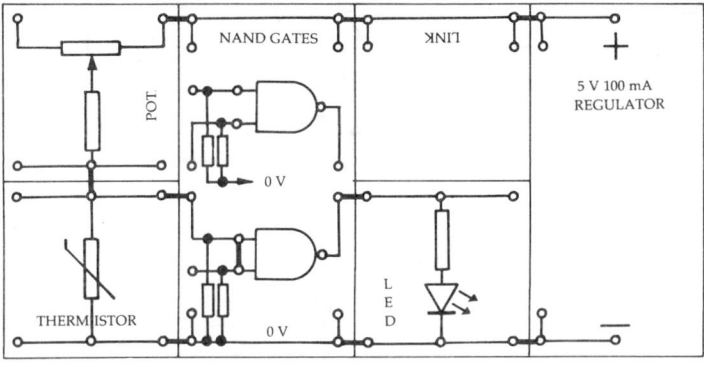

Circuit 30(a)

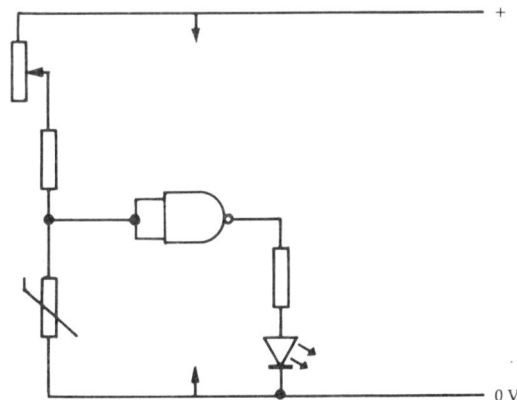

Circuit 30(a) causes the LED to glow when the temperature of the thermistor rises. The pot. must therefore be set so that the LED is just off. When the thermistor is touched or breathed on, the LED should glow.

Circuit 30(b) does the opposite, and should be set so that the LED is just glowing before the thermistor is warmed up.

The theory behind the circuit is very similar to that of the previous investigation, but is described again for convenience. In Circuit 30(a) the pot. (connected as a variable resistor) and thermistor are wired to form a potential divider, with their junction connected to the inputs of a NAND gate wired as an inverter. If the temperature of the thermistor rises, its resistance falls, causing the voltage at the junction to decrease. This 'low' causes the inverter's output to go high, making the LED light.

In Circuit 30(b) the thermistor and pot. are interchanged. As the temperature of the thermistor rises, the voltage at the junction between the thermistor and pot. rises, causing the NOT gate output to go low.

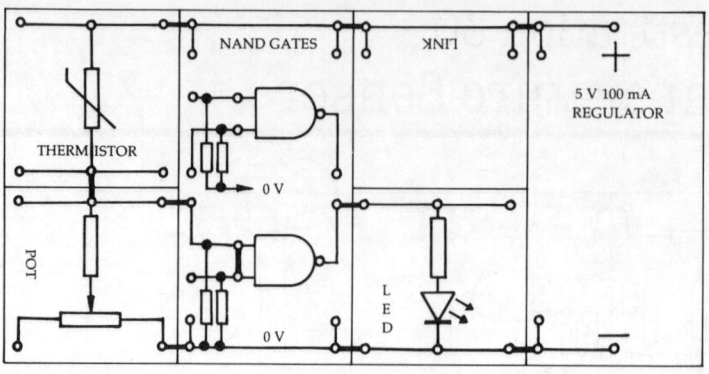

Circuit 30(b)

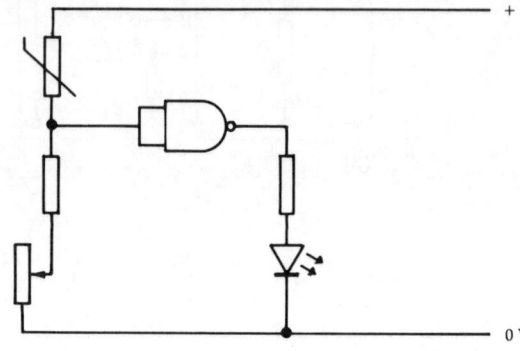

Further Work   A thermistor probe could be used with liquids as described in
Investigation 17. Pupils could use this probe, with Circuit 30(b), to
monitor the temperature of some water in a beaker. A transistor, relay
and buzzer could then be connected to produce a warning device if the
temperature falls below a pre-set level.

More advanced pupils could manufacture a small heater, made by
winding a coil of nichrome wire round a folded wooden splint. It is
important to check that the current flowing through the coil is less than
1 A at, say, 12 V. In other words the resistance of the wire should be
greater than 12 Ω. (The teacher could first ascertain the length of
nichrome wire required to achieve this.) The heating coil should then be
connected to an independent power supply (*not* the 5 V regulator) via
the relay. 'Independent' could mean a completely separate power unit,
or the power unit supplying the regulator. Assuming that the latter is
set to 12 V, it may well be the most convenient power source, and the
complete circuit is shown at the top of page 105.

The heating coil should be placed on a ceramic tile (or similar non-
metallic, non-combustible material) and the thermistor positioned
carefully above it. The circuit should automatically switch the heater on
and off, thus maintaining the temperature of the heater at a fairly
constant level.

Pupils should be instructed not to let the relay 'chatter'. If the pot. is
set so that the heater becomes quite hot (without burning the wooden
splint), the circuit should react quickly enough to achieve a fairly 'clean'

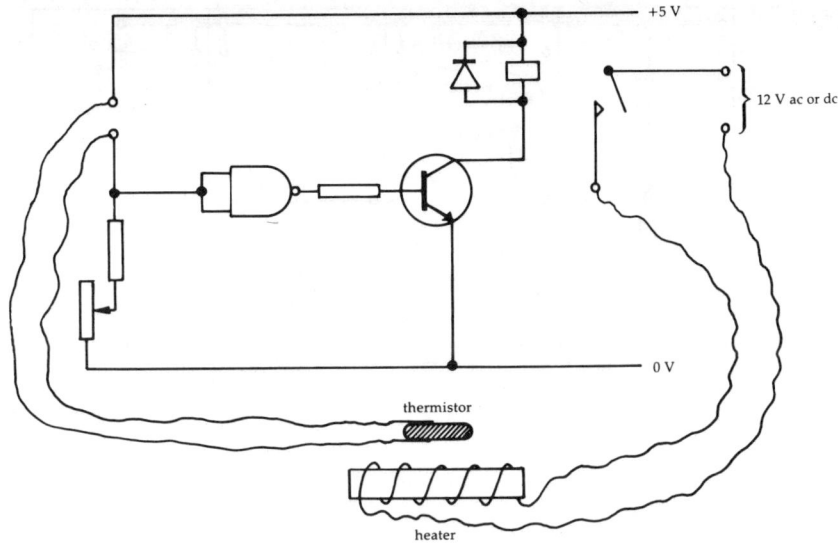

switch-over between on and off. The contacts of a 'chattering' relay may quickly burn out with a current of 1 A flowing.

A second, more powerful relay could be used to avoid this problem. The heater could then be placed in distilled water if preferred, with the thermistor placed at the top of the water. The thermistor would need to be well insulated if ordinary tap water were used. A thermometer could then be used to check the stability of the temperature achieved.

Homework
Suggestion

Pupils could be asked to design a fire alarm based on the NOT gate circuit. Alternatively, fridge and freezer failure alarms are popular gadgets which many pupils might find interesting.

Investigation 31
Simple Latch Switch

When the push switch is pressed, the output should go high, as shown by LED B glowing. This 'high' is fed back to one of the upper gate inputs keeping the circuit 'latched' even when the push switch is released.

LED A indicates when the push switch is pressed. It can be omitted if preferred.

The inputs to the lower NOR gate are connected together to produce

105

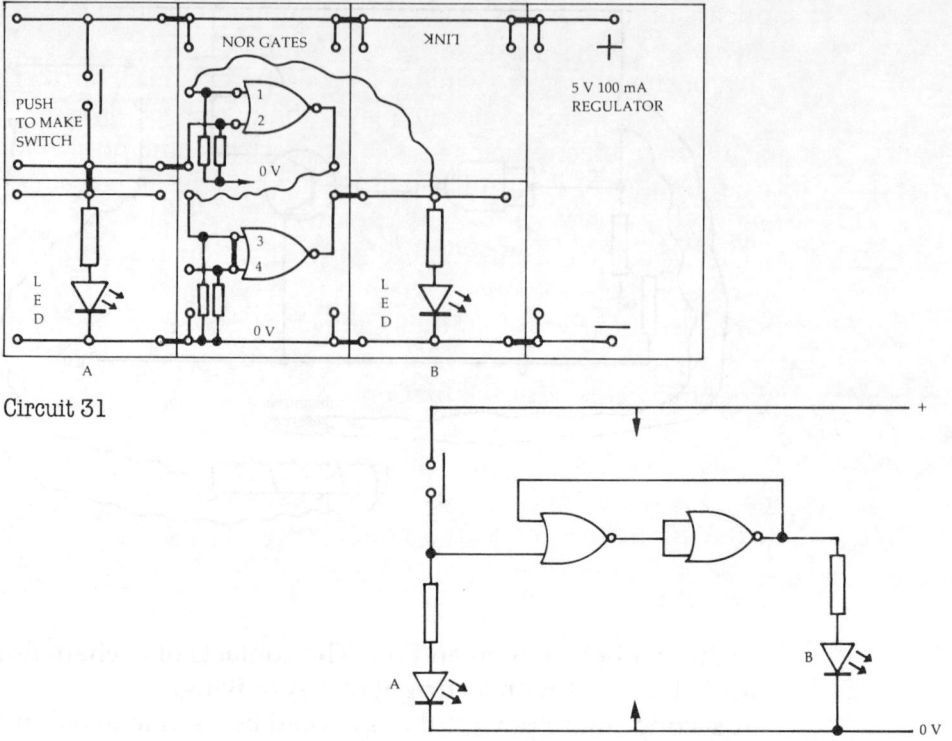

Circuit 31

a NOT gate. This inverts the action of the upper NOR gate, producing the same result as an OR gate, as shown below.

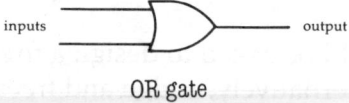

OR gate

When the output of an OR gate is connected to one of its inputs, the gate can be latched with its output high by making its other input high for an instant.

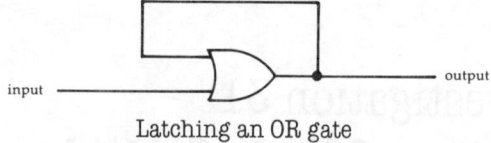

Latching an OR gate

The gate can be reset (unlatched) by switching off the power supply, though it may be necessary to wait for several seconds until the capacitors across the power supply discharge. This delay may be avoided by removing the link connecting the regulator to the link board. Alternatively, the circuit can be reset by momentarily disconnecting the feedback wire. (*See also* 'Further Work'.)

Applications   Electronic latches are widely used in electronics. For example, when a calculator key is pressed, the number represented is 'latched' and appears on the display. Two sets of numbers have to be latched before they can be processed.

An ordinary toggle switch is a type of mechanical latch. The advantage of a electronic latch is its ability to latch or reset (clear) by means of an electronic message. An electronic memory consists of a large number of electronic latch circuits which can be set or reset as required.

The type of circuit used in this investigation also lends itself to alarm systems, where the push switch is replaced by a 'pressure mat'. This is placed under a carpet, and its contacts close when the mat is stepped on. Several pressure mats can be connected, providing that they are wired in *parallel* with the first one.

Further Work

Pupils could be asked to use a second push switch so that the circuit may be unlatched (i.e. reset) as required. There are several ways of achieving this. Some pupils may produce one of the following circuits:

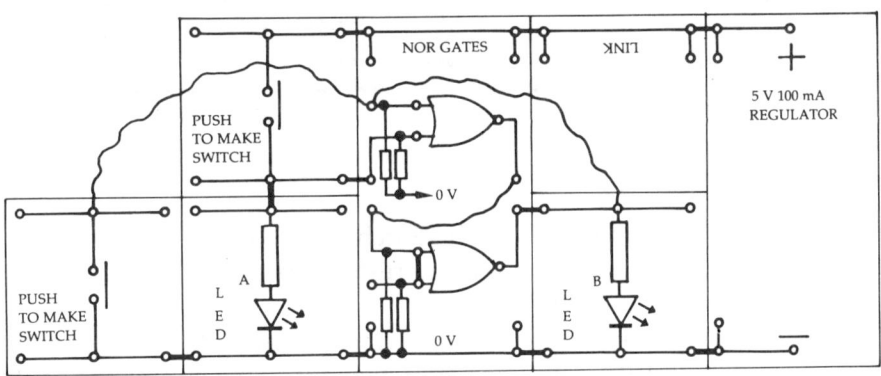

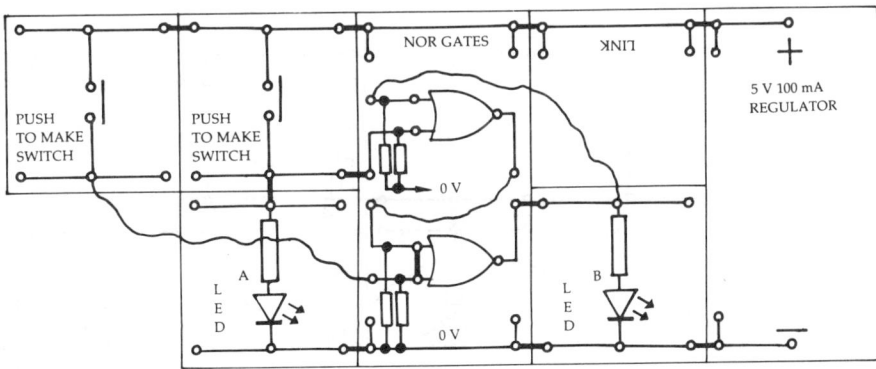

The problem with both circuits is that when the reset switch is pressed, it effectively short-circuits the output from one or other gate. The fact that this does not damage the gate is due to its excellent

107

electrical properties, rather than the intelligence behind this method.
A better method would include a resistor, as shown below.

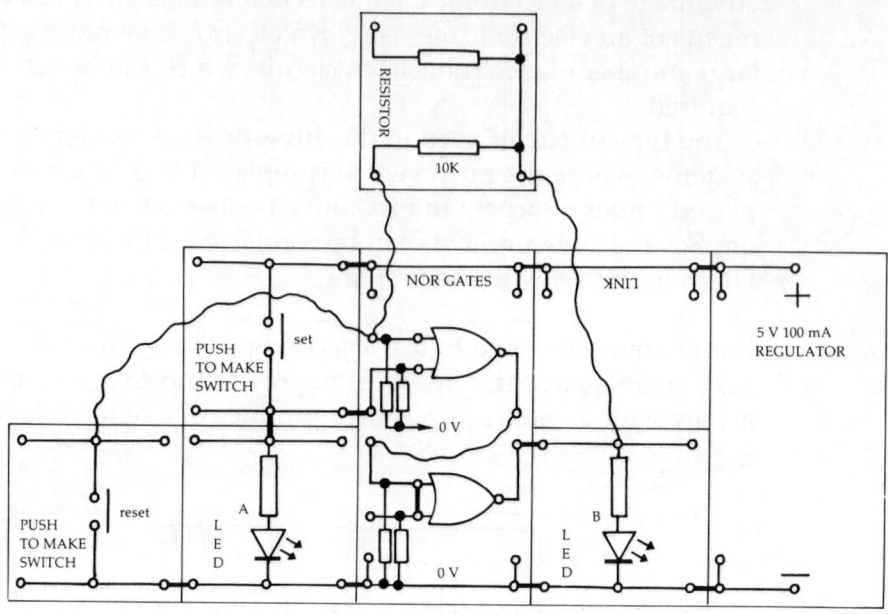

When the reset push switch is pressed in this circuit, a very small current flows through the resistor as the voltage at the input falls to zero.

Another alternative is to remove the link which connects the lower gate inputs. The reset switch can then be connected to the spare input, as shown below.

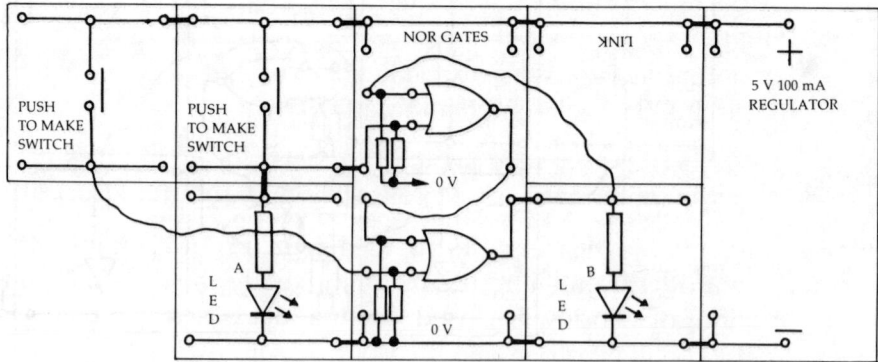

<table>
<tr><td>Homework
Suggestion</td><td>Pupils could be asked to design a circuit capable of storing four or more binary digits which are displayed by means of LEDs. More advanced pupils may be able to devise a method by which a single reset switch</td></tr>
</table>

could reset all the latch circuits at the same time. A possible solution is given below.

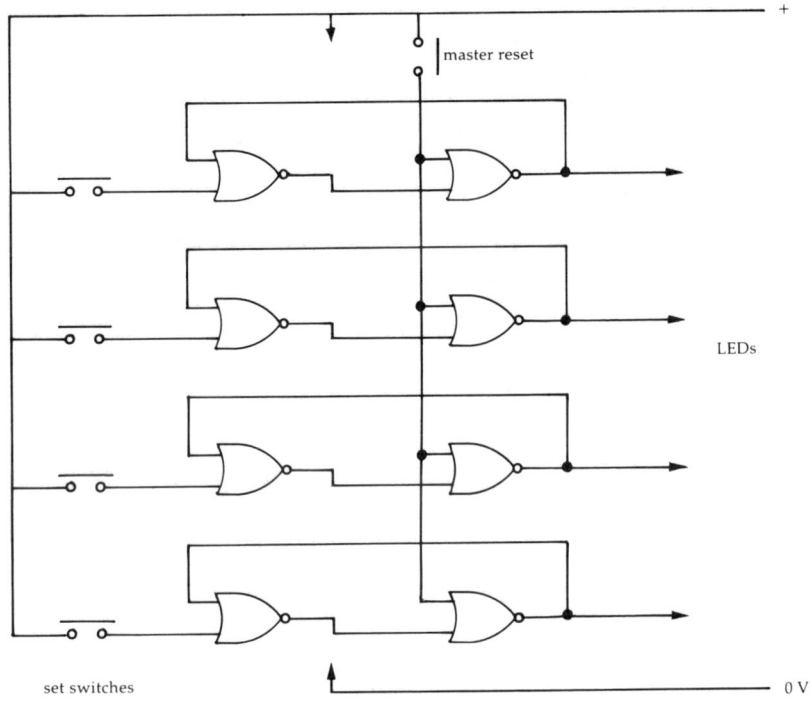

Investigation 32
Reed Switch Alarm

Circuit 32 shows how a reed switch may be connected to activate an LED when a magnet is moved *away* from the switch. Pupils should take note that this is the opposite effect to that obtained with the previous circuit. The following summary may help:

Magnet near reed switch; contacts closed; input 'low'; output 'low'.
Magnet moved away; contacts open; input 'high'; output 'high'.

In a practical circuit, the reed switch could be mounted in a door or window frame, and the magnet fixed to the door or window. When the door or window is closed, the magnet is near the reed switch, and the LED will be off. When the door or window is opened, the magnet moves away from the reed switch, and the LED switches on. In practice, a buzzer (or siren) would be used instead of the LED. Circuit 34(b) on page 115 shows how a buzzer may be connected.

The lower NOR gate inverts the effect of the upper gate, producing the action of an OR gate. Since the 'spare' input of the upper gate is

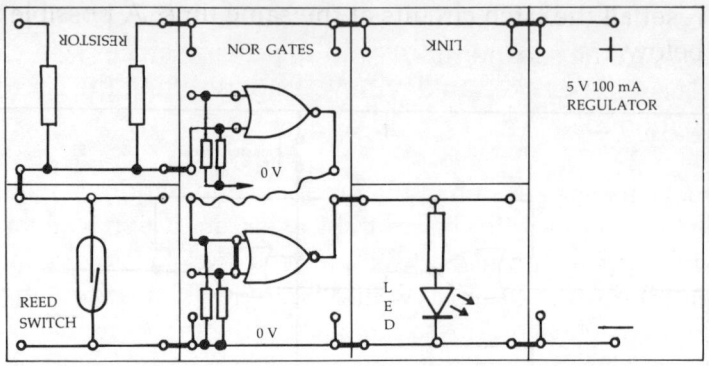

Circuit 32

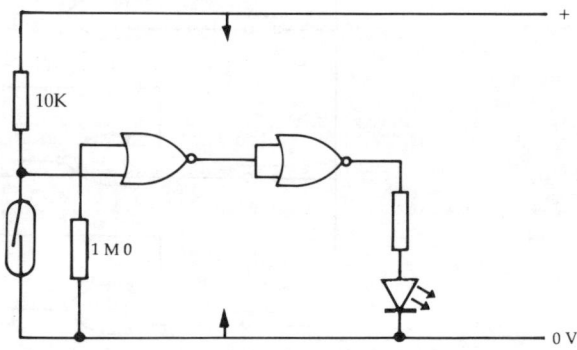

held 'low' by means of the 1 MΩ resistor mounted on the NOR board, the OR gate effectively functions as a buffer. When the magnet is placed next to the reed switch, the contacts close and the buffer's input is connected to zero volts. When the magnet is moved away, the reed switch contacts open, and the resistor causes the voltage at the buffer's input to rise to logic 1. This in turn causes the output to go high, switching on the LED.

Further Work

At present, the intruder could simply close the door or window to silence the alarm shown in Circuit 32. A latch is therefore required so that, once triggered, the alarm continues to sound. There are several ways of achieving this. One method is illustrated below.

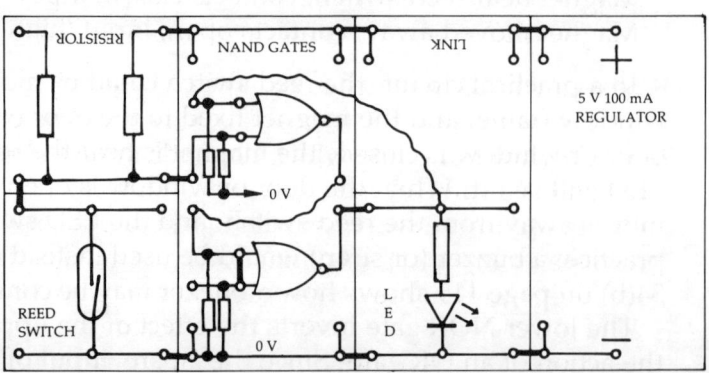

The circuit works as before, except that once the output goes high, it feeds back to the second input, and the circuit latches.

Pupils could be asked to devise ways of resetting the circuit (as in Investigation 31).

In a practical alarm system, several doors and windows would probably have to be included. Pupils could be asked to design a suitable system. The solution is to connect additional reed switches in *series* with the first one. An interesting exercise would be to devise a system incorporating both reed switches and pressure mats in the same circuit. There are many ways of achieving this, but a possible solution is to place the pressure mat (represented by a push switch in the Kit) in parallel with the 10 kΩ resistor. An additional 1 kΩ resistor (not included in the Kit) is required to prevent a short circuit when the switch is pressed.

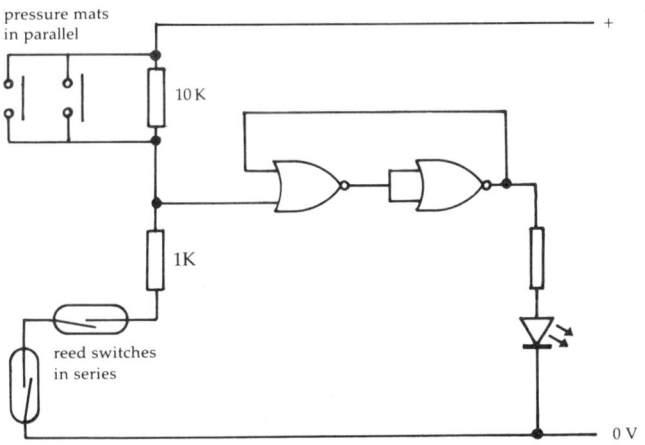

Investigation 33
Capacitor Delay

When Circuit 33 is switched on, the LED may glow for a time. After the LED has switched off, the push switch should be pressed. This makes the LED glow for a time which depends upon the resistance (of the pot.) and the value of the capacitor.

When the push switch is pressed, the inputs of the NAND gate (wired as an inverter) are connected to zero volts. Its output therefore goes high. Any charge on the capacitor is removed via the push switch.

When the switch is released, the discharged capacitor holds the input at zero volts. Current flows through the pot. (connected as a variable

111

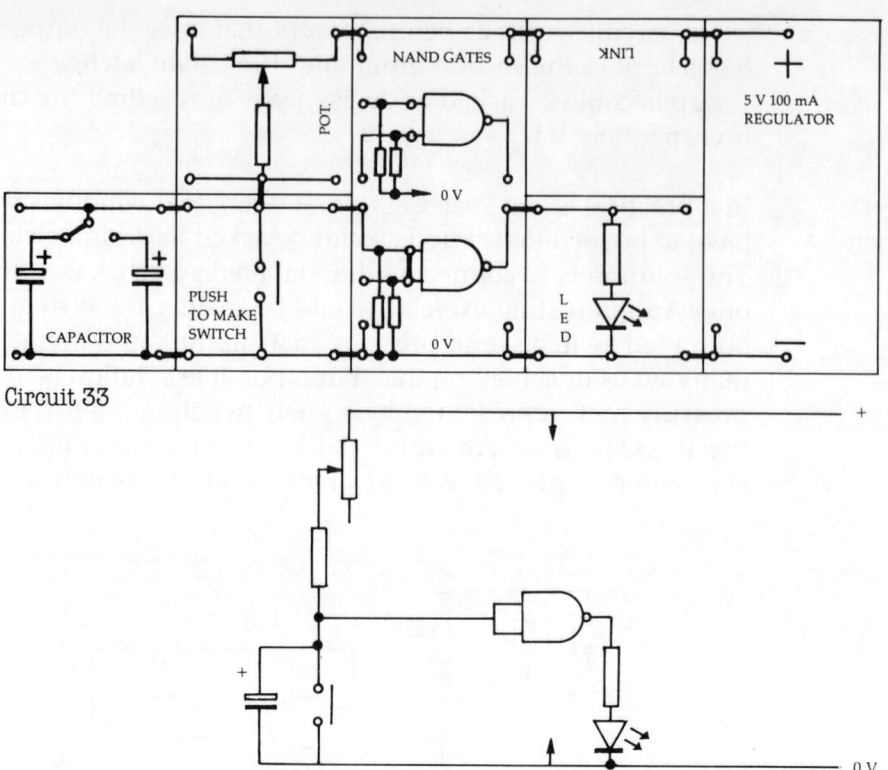

Circuit 33

resistor) at a rate determined by its setting. The capacitor charges, and when the voltage at the NAND inputs has risen sufficiently, the output goes low, causing the LED to stop glowing.

Background Information

The approximate time for which the LED glows can be determined as follows:

Time (seconds) = Resistance (ohms) × Capacitance (farads)
or
Time (seconds) = Resistance (megohms) × Capacitance (microfarads)

This can be checked by turning the pot. fully anti-clockwise, making its value 50 kΩ. The value of the series resistor is small by comparison, and may be disregarded. Assuming that the smaller capacitor is connected, the time for which the LED glows is as follows:

Time (seconds) = 0.05 (megohms) × 22 (microfarads)
= 1.1 seconds

By similar reasoning, the maximum delay with the larger capacitor is about 50 seconds. It should be noted, however, that the accuracy of electrolytic capacitors is poor, and different capacitors (of the same nominal value) may produce different results.

Further Work

The LED may be replaced by a transistor/relay/ buzzer as shown in Investigation 34. Pupils may note that having a buzzer sounding while

112

they are timing, and stopping at the end of the timed period, is the wrong way round. Ideally the buzzer should be silent during the timed period, and should sound at the end of the period.

There are several ways of achieving this, and some pupils may discover them. Possible solutions are as follows:

1 Both NAND gates (connected as inverters) could be used in series, so that the total effect is that of a buffer.
2 The buzzer could be connected to the 'normally closed' relay contact.
3 The output could be connected as a 'sink' (shown below). This will only work with the LED, since the transistor supplied will not operate with the output as a sink. However, some pupils may try to replace the LED directly by the buzzer. While this should not work properly, in practice it usually does!

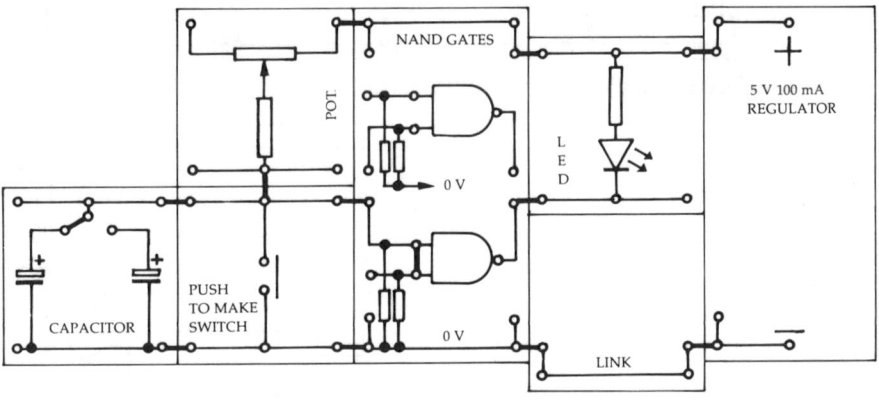

Homework Suggestion Pupils could be asked to design a timer, using the circuit with a buzzer, and a double-pole switch. One pole of the switch should connect the circuit to the positive side of a battery (i.e. a normal on/off switch). The second pole should be wired in place of the push switch, so that when the timer is switched on, the second pole is open, and when switched off the second pole discharges the capacitor. Such a circuit could become a useful timer. A 4.5 V battery may be employed if desired.

Investigation 34
Driving a Transistor (and Relay)

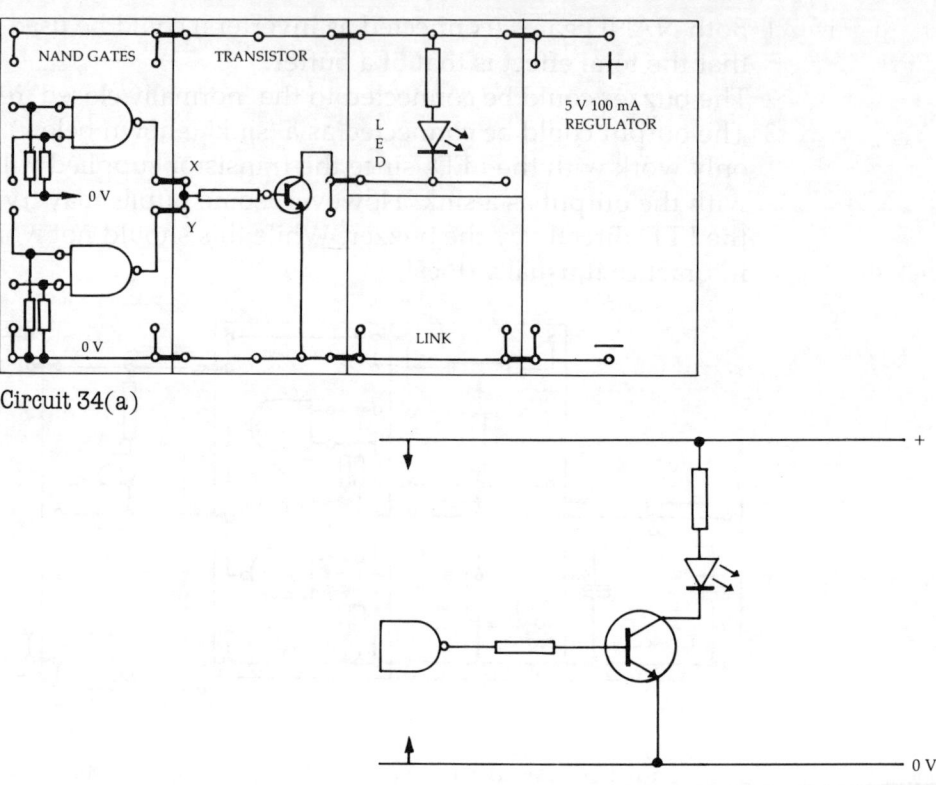

Circuit 34(a)

Circuit 34(a) is not a complete circuit, but shows how a transistor may be connected to the output of either gate. The transistor may then drive an LED, or a more powerful device such as the relay.

Any of the previous circuits may be used with the transistor, but care should be taken to fit the link marked 'X' *or* 'Y' in the diagram above. *Never* fit a link in both positions 'X' and 'Y' as this will allow a large current to flow between the gate outputs if one is high and the other is low. The outputs are designed to withstand this current, but it is still best to avoid it.

Background
Information

A current of only a few milliamps is available at the outputs of the gates. Whilst this is excellent compared with previous ranges of ICs such as the orginal 74TTL, 74LS TTL and CMOS4000 families, it is insufficient (in theory) to operate the buzzer or relay.

However, the current can be fed into the base of a transistor via the base current limiting resistor. This enables the transistor to switch on a much larger current via its collector and emitter.

114

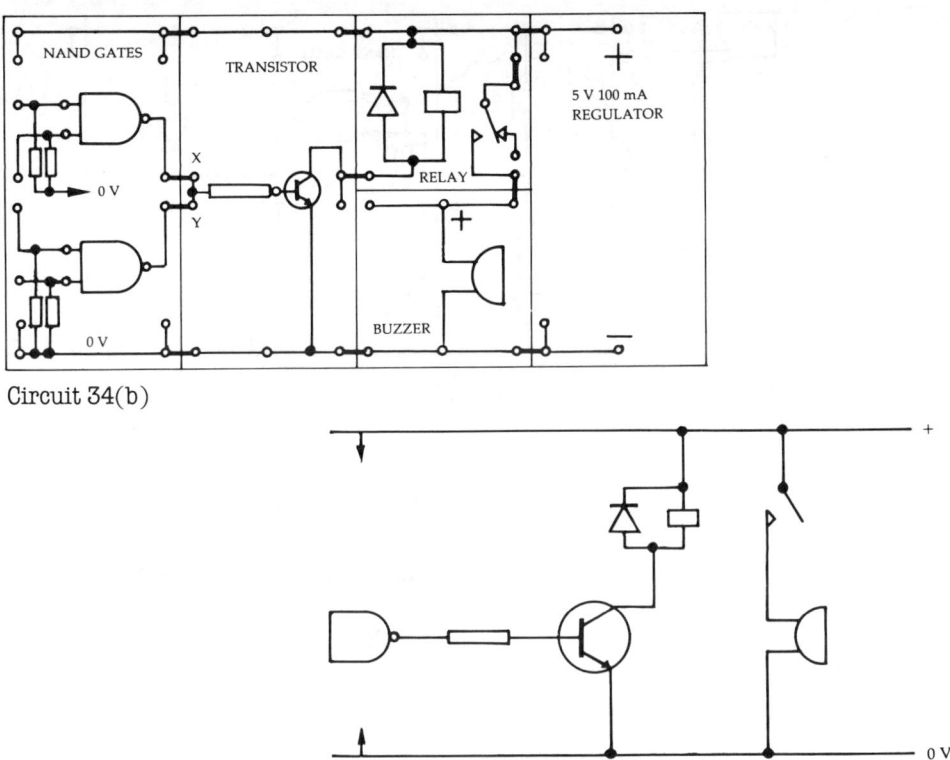

Circuit 34(b)

Circuit 34(b) shows how a relay may be connected to the previous circuit to operate a buzzer reliably, or any other equipment within the limits imposed by the relay contacts (*see* Investigation 6).

Investigation 35
Energy Saving Circuit Using a NAND Gate

When adjusting Circuit 35 it is essential to hold down the push switch, and avoid shading the LDR by mistake. The pot. must be set so that the LED is just off.

   When the LDR is in darkness, the push switch will operate the LED, but when the LDR is exposed to light, the LED will not operate. The LED simulates a porch or room lamp designed to operate only at night. The push switch represents a toggle switch. If left on accidently, the lamp will automatically switch off in daylight.

Circuit Theory   The upper NAND gate is used, its output being connected to the LED as a *sink*. When the output is high the LED is off; when the output is low, the LED glows.

115

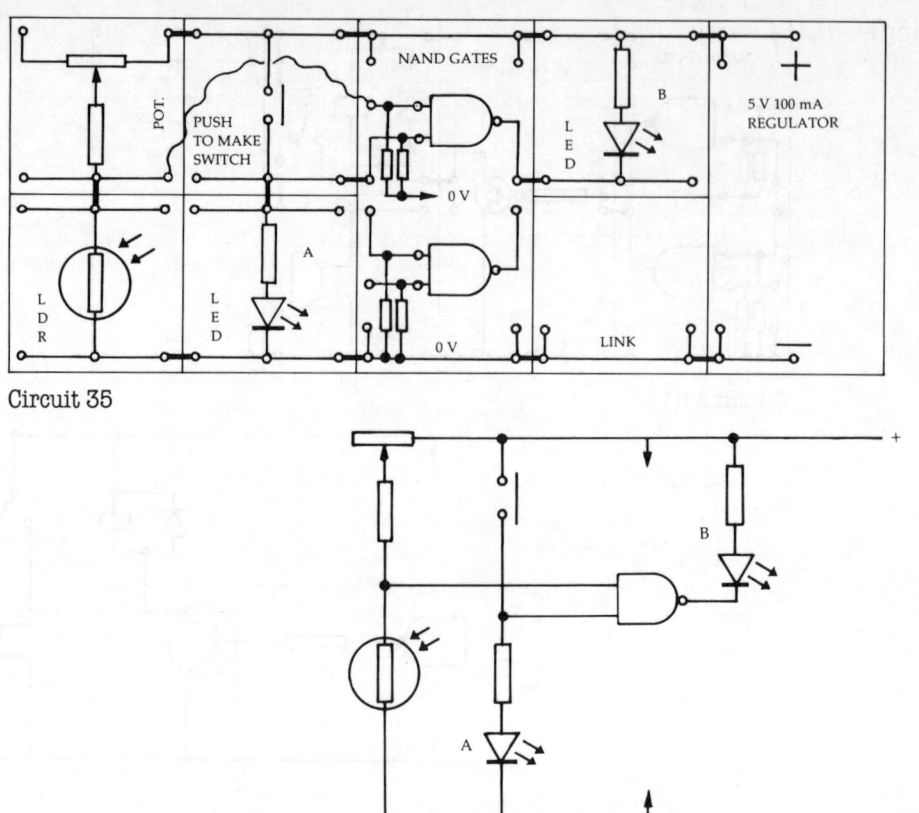

Circuit 35

A glance at a NAND gate truth table will reveal that the output is at logic 0 only if *both* inputs are at logic 1. Any other combination of logic states at the input will produce a logic 1 at the output.

Both inputs of upper gate must therefore be high, in order to cause LED B to light. One is connected to the junction between the pot. (wired as a variable resistor) and the LDR; the other is connected to positive via the push switch. LED A is only included to indicate when the push switch is pressed.

Assuming that the pot. is adjusted correctly, when the LDR is exposed to light its resistance is sufficiently low to produce a logic 0 at the upper gate input. Thus LED B remains off regardless of the state of the other upper gate input. When the light falling on the LDR is reduced, its resistance rises, causing the voltage at its junction with the pot. to rise to logic 1. LED B still remains off, but when the push switch is pressed, *both* upper gate inputs will be at logic 1, and LED B will glow.

Further Work Pupils could be asked to use the output as a *source*, which would then enable the transistor and relay to be employed. However, the effect produced will be the opposite of that required. Some pupils may be able to deduce that if the lower NAND gate is connected as an inverter in

116

series with the upper gate, and the lower gate output is used as the source, the circuit will function as before.

A further saving in energy could be achieved if the LED glowed for a limited time after pressing the push switch. Investigation 33 showed how an LED could be made to glow for a limited time. Pupils could be asked to combine the two circuits so that the LED glows for a few seconds after pressing the push switch *only* when the LDR is in darkness.

   There are several ways of achieving this. Some pupils may connect the capacitor delay circuit to a relay. The relay contacts could then be used as the push switch in the second circuit. A more elegant solution would use the NAND gate to combine the requirements. Pupils may need to use a second pot. if they wish to test their designs. However, the pot. in the delay circuit could be replaced with the 10 kΩ resistor to provide a fixed delay, leaving the pot. available to be placed in series with the LDR. A possible solution would then be as shown below.

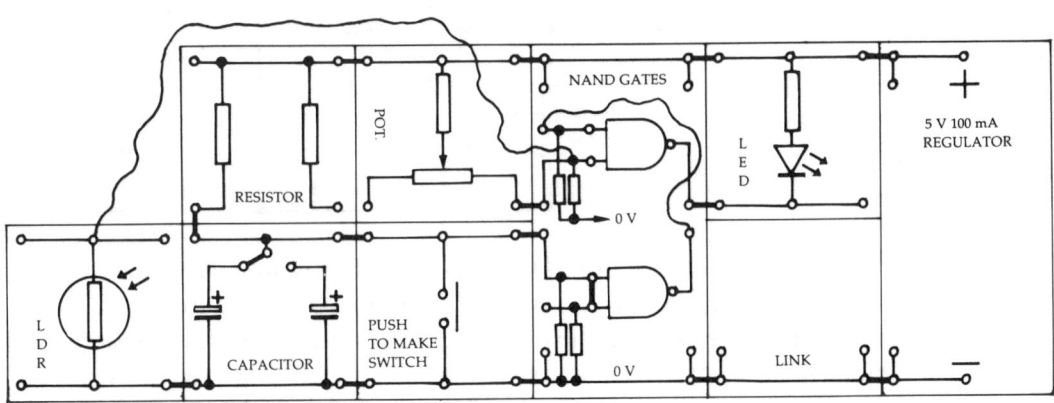

Investigation 36
Greenhouse Plant Waterer

In Circuit 36 the pot. should be adjusted so that the LED just stops glowing when the LDR is exposed to light and the moisture sensor probes are apart (or in dry soil). When the LDR is shaded the LED should glow, indicating that the water valve or pump is switched on. If the moisture sensor probes are in damp soil (or twisted together) the LED should switch off.

The junction between the pot. (connected as a variable resistor) and LDR is connected to the first NOR gate, which is wired as an inverter.

117

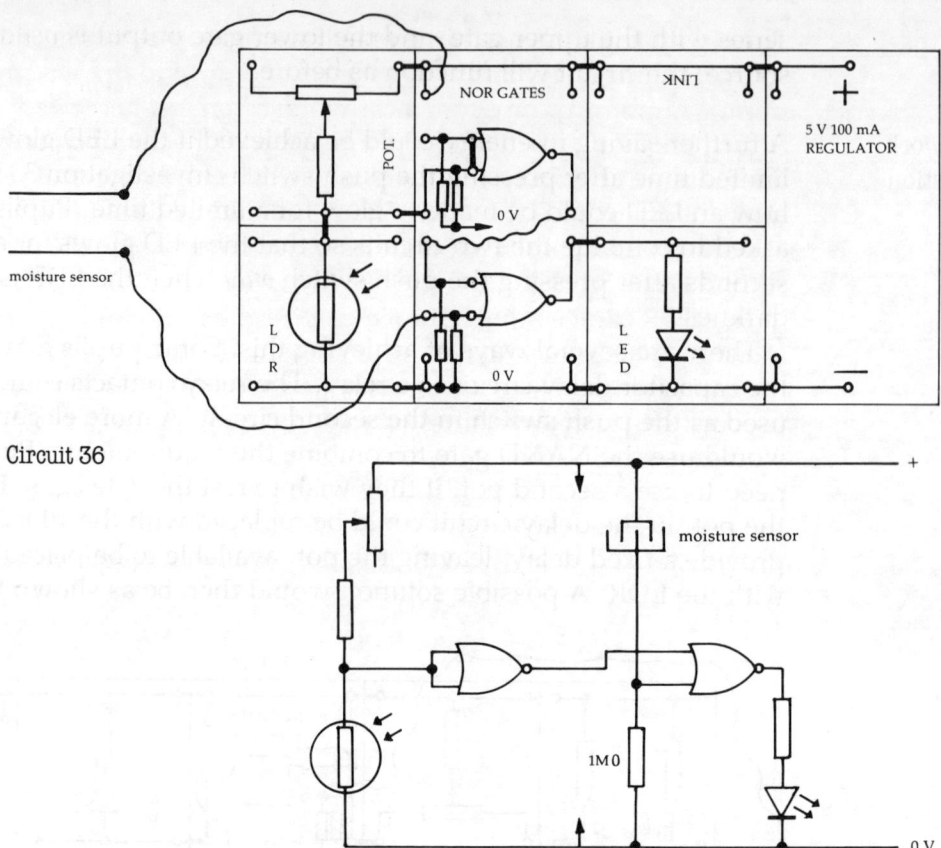

Circuit 36

As the light level on the LDR decreases, the voltage at the inverter's input rises. The inverter's output therefore becomes logic 0. This is connected to one of the inputs of the second NOR gate. Water on the moisture sensor would cause the other output to go high. However, in dry conditions the voltage on this input falls to near zero, due to the 1 MΩ resistor mounted on the NOR gates board. When both inputs of the NOR gate are at near zero volts, its output goes high, lighting the LED.

The circuit waters the plants at night when the soil is dry. If the pot. and LDR are interchanged as shown below, the circuit will water the plants in daylight when the soil is dry.

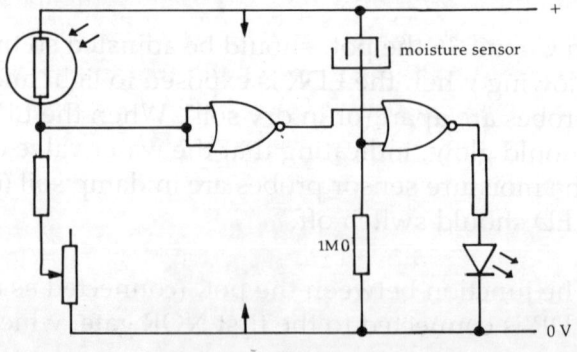

The circuit could be used in a practical situation by connecting a transistor and relay, and using the relay output to drive a small electric pump. A motor/pump from a car windscreen washer (possibly obtained from a breaker's yard) could be employed. Check the current used by the motor; if it is more than 1 A, a second, more powerful relay must also be employed. The relay contacts should be disconnected from the 5 V supply rails, and a separate 12 V supply used for the motor, as shown on page 33.

The LDR can be adjusted by means of the pot. However, the moisture sensor has no means of adjustment and may allow the soil to become too dry. The problem may be overcome by positioning the probes further apart in the soil, making them shorter, or placing them at the top of the soil.

Investigation 37
House Alarm

To operate Circuit 37, pupils should check that the magnet is in place against the reed switch, and adjust the pot. until the buzzer switches off. The buzzer should start again when the magnet is moved away from the reed switch, or when the push switch is pressed. Alternatively, the buzzer should sound if the thermistor is warmed up, providing that the pot. is set accurately.

Circuit Theory

The push switch represents a pressure mat, normally placed under a carpet. The switch is 'normally open' (i.e. off), and closes (switches on) when stepped on. The reed switch is 'normally closed' when the magnet is in place – as it would be when a door or window is closed. When the door or window is opened the magnet moves away, and the reed switch opens.

A more complex circuit employing several gates is best described in the form of a truth table, remembering that:

Push switch not pressed	a = logic 0
Push switch pressed	a = logic 1
Magnet near reed switch	d = logic 1
Magnet removed from reed switch	d = logic 0
Thermistor cold	f = logic 0
Thermistor hot	f = logic 1

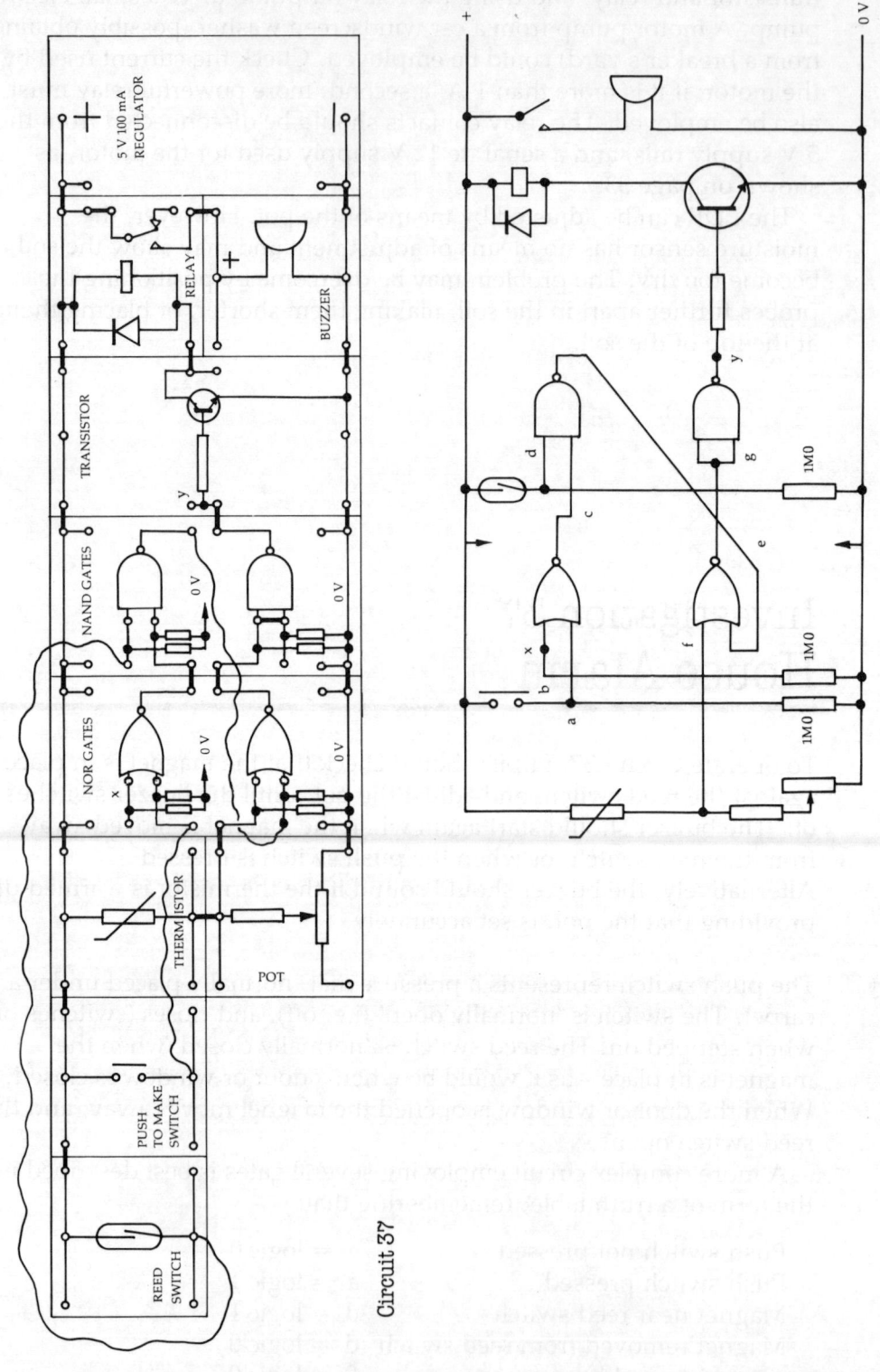

Circuit 37

The following sequence is best understood by consulting the truth tables for NOR and NAND gates.

a	b	c	d	e	f	g	Output

With push switch open, reed switch closed, thermistor cold:

a	b	c	d	e	f	g	Output
0	0	1	1	0	0	1	0

With push switch closed, reed switch closed, thermistor cold:

a	b	c	d	e	f	g	Output
1	0	0	1	1	0	0	1

With push switch open, reed switch open, thermistor cold:

a	b	c	d	e	f	g	Output
0	0	1	0	1	0	0	1

With push switch open, reed switch closed, thermistor hot:

a	b	c	d	e	f	g	Output
0	0	1	1	0	1	0	1

Notice that the output is at logic 1 when either the push switch is pressed, or the reed switch opens, or the thermsistor is hot. If all three events occur together, the truth table is as follows:

a	b	c	d	e	f	g	Output
1	0	0	0	1	1	0	1

Problem Solution In the sequences above, the input labelled 'b' is always low, due to the 1 MΩ resistor. If input b is connected to the output (by connecting points x and y together) the circuit will latch with the output high, even if the conditions return to normal:

a	b	c	d	e	f	g	Output
0	1	0	1	1	0	0	1

The output is connected to a transistor, relay and buzzer, as explained in previous investigations.

Further Work/Homework A house alarm system would normally employ a number of reed switches, and pressure mat switches. Pupils could be asked to work out how these extra switches should be connected.

Extra reed switches must be connected in *series* with the first one; extra push switches must be connected in *parallel*.

Investigation 38
Quiz Show Reaction Detector

When Circuit 38 is switched on, neither LED should light. If button P is pressed first, LED p glows, and button Q becomes inoperative. If button Q is pressed first, LED q glows, and button P becomes inoperative.

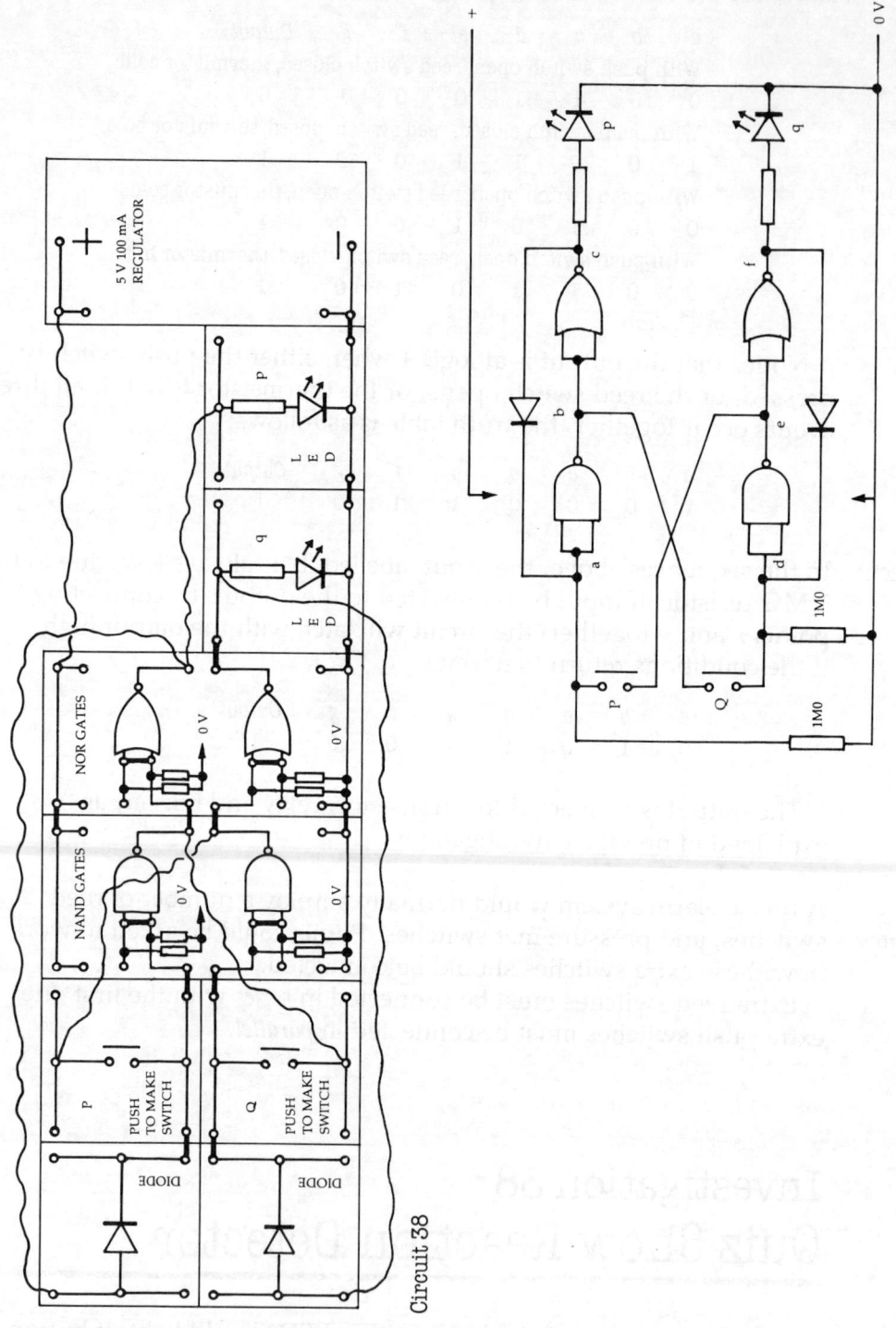

Circuit 38

The circuit can be used for quiz games, or as a reaction tester between two contestants.

The four gates are all connected as inverters. When first switched on, they assume the following states:

a	b	c	d	e	f
0	1	0	0	1	0

Note that switch P is operative, as it is connected to 'e' which is high. By the same reasoning, switch Q is operative, as 'b' is high.

If the switch P is pressed, 'a' goes high, as shown:

a	b	c	d	e	f
1	0	1	0	1	0

Thus LED p glows. However, since 'b' is now low, switch Q is no longer operative. The upper diode causes 'a' to remain high after switch P is released.

The circuit may be reset by switching off the power supply, then switching it on again. If switch Q is now pressed, 'd' goes high:

a	b	c	d	e	f
0	1	0	1	0	1

LED q now glows, but as 'e' is now low, switch P is no longer operative. The lower diode causes 'd' to remain high after switch Q is released.

The two diodes have been used as they are readily available in the Kent Electronics Kit. Two separate resistors (say, 10 kΩ) could be substituted, in which case all the 1 MΩ resistors already fitted to the boards would be superfluous, as every input would then be connected to an output, either directly or via a resistor. The diagram below shows the resulting circuit.

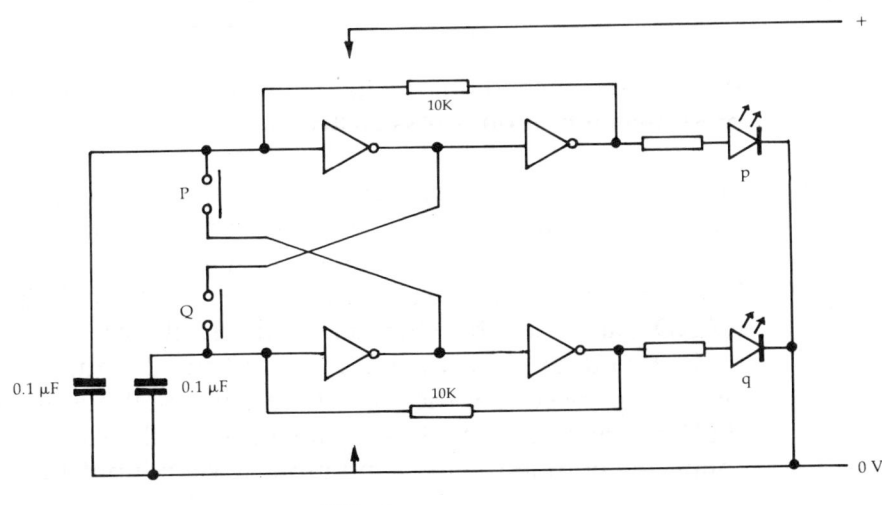

Note the addition of the two capacitors (not included in the Kit) with values around 0.1 μF. These ensure that the circuit never latches accidently – especially when first switched on.

Investigation 39
Bistable Multivibrator

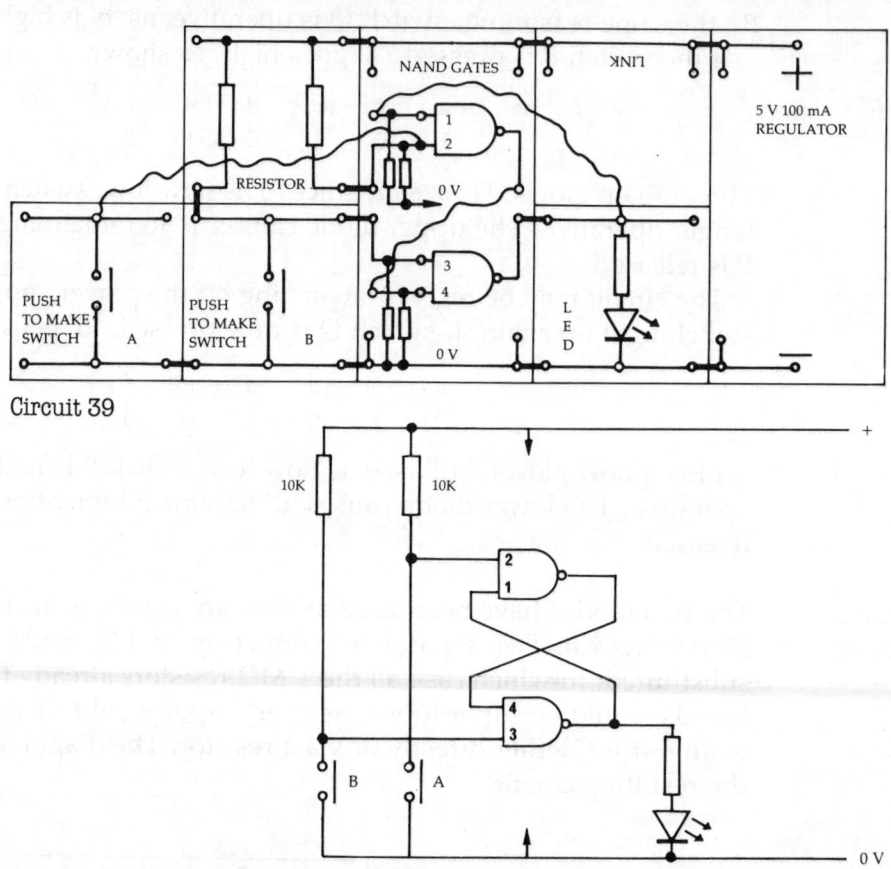

Circuit 39

When first switched on, the LED in Circuit 39 may be on or off. If off, press switch B, if on, press switch A.

Circuit Theory

The *bistable* has two stable states, and may be latched in either state by means of switches or other electronic circuits. It may be likened to a toggle switch, which latches in one position or the other.
Understanding this type of circuit may prove difficult, but viewing a NAND gate truth table should provide assistance.

The two 10 kΩ resistors hold inputs 2 and 3 high, unless either push switch is pressed. For the moment, assume that the circuit is latched off (LED not glowing). The lower gate output will be low, and this is connected both to the LED and input 1. With input 1 low and input 2

124

high, the upper output is high. This is connected to input 4, and with both inputs 3 and 4 high, the lower ouput remains low. The circuit is stable in this state.

When switch B is pressed (and held), input 3 goes low (since it is connected to zero volts via the switch). The lower gate changes state, its output going high. The LED lights, and input 1 becomes high. With both inputs to the upper gate being high, its output goes low. This causes input 4 to go low. When switch B is released, although input 3 goes high again, the lower gate does not change state as input 4 is low. This sequence happens so quickly that the slightest touch on switch B will cause the circuit to latch on.

When switch A is pressed (and held), input 2 is made low and the upper gate changes state. Its output, which is now high, makes input 4 go high. With both inputs 3 and 4 high, the lower output goes low and the LED switches off. Input 1 is now low, and this maintains the upper gate in the new state even when switch A is released. The circuit is now latched with the LED off.

Background Information

The four 1 MΩ resistors mounted on the NAND gates board serve no useful purpose in this circuit and should be disregarded. The two 10 kΩ resistors are so much lower in value than 1 MΩ that inputs 2 and 3 are held high unless one of the switches is pressed.

Applications

The bistable is a very important circuit in digital electronics. A typical application is when it is used to 'latch on' the play mode of a video recorder after receiving a signal from a remote-control device. It is used extensively as an 'debounce' circuit. Push button switches and keys tend to produce several pulses as they are pressed, rather than one 'clean' transition from logic 0 to logic 1. If connected via a bistable, the bistable output will change cleanly from one logic state to another, regardless of any bouncing of the switch contacts.

Further Work

This type of bistable can be used to debounce a two-way toggle switch, particularly if the switch is intended for use with other digital circuits,

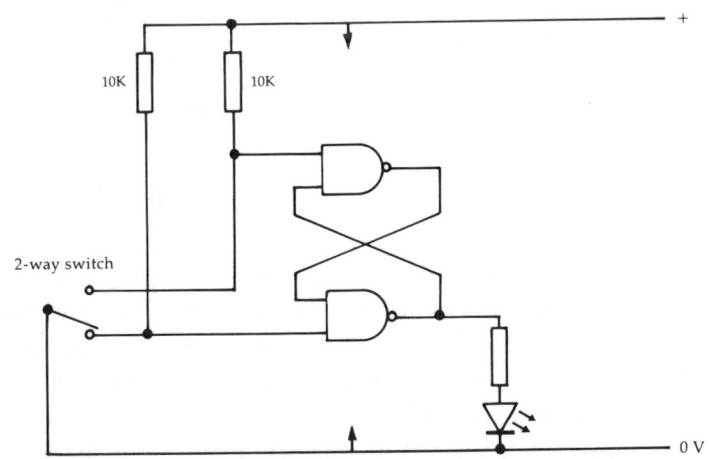

such as counters. Such circuits require very 'clean' transitions between logic 0 and logic 1. The push switches shown in Circuit 39 could be replaced with a two-way toggle switch, as shown on page 125.

Homework
Suggestion

Latching alarm circuits have already been discussed. Pupils could be asked to turn this latching circuit into an alarm system designed to operate with under-carpet pressure mats. Pupils may be able to deduce that switch B must become the pressure mat (or several in parallel), and switch A is the reset switch.

Investigation 40
Monostable Multivibrator

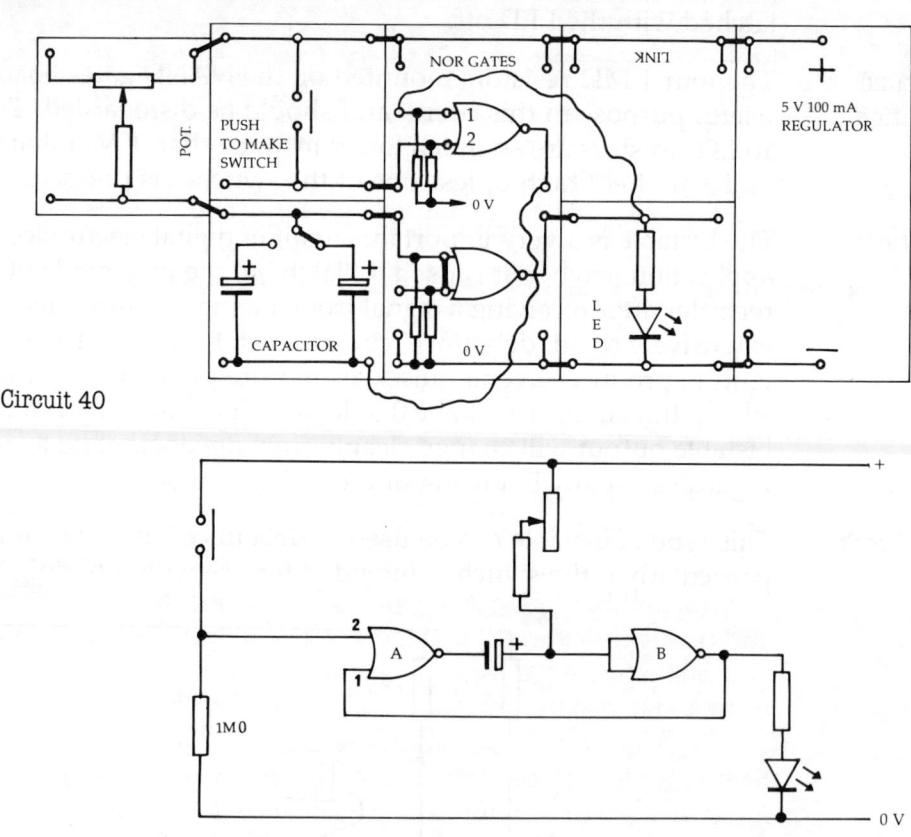

Circuit 40

When the push switch in Circuit 40 is pressed, the LED should light for a time determined by the setting of the pot. and the value of the capacitor. Pupils should note that the time for which the push switch is held down makes no difference to the time for which the LED lights. If the push switch is held down for longer than the 'timed period' of the monostable, the result should be the same as before. However, switch

126

contact bounce may cause the monostable to retrigger as the switch is released.

The *monostable* has only one stable state. It may be changed into its other state (by pressing the switch), but returns to its stable state after a time delay.

A truth table may clarify the method of working.

When first switched on, gate B's inputs will rise to logic 1 since they are connected to the positive rail via the pot. The logic states will be:

Inputs gate A		Output gate A	Inputs gate B	Output gate B
1	2			
0	0	1	1	0

When the switch is pressed, pin 2 will become logic 1. Gate A's output will rapidly fall to logic 0, and the falling potential will be transferred across the capacitor, making the input of gate B logic 0. Its output therefore switches to logic 1, causing input 1 to become logic 1:

1	1	0	0	1

If the push switch is then released, the situation hardly changes:

1	0	0	0	1

A potential difference is present across the pot. and current flows, charging up the capacitor, eventually causing gate B's inputs to rise to logic 1. Its output therefore switches to logic 0, causing input 1 to switch to logic 0:

0	0	1	1	0

Now that both gate A's inputs are low, its output changes to high, and this is transferred across the capacitor reinforcing the rising potential at gate B's inputs. The monostable has now returned to its former stable state.

Some pupils may be able to discover the relationship between the resistance (R) of the pot., the capacitance (C) of the capacitor and the time period (T). The approximate relationship is:

$$T = R \times C$$

where T is in seconds, R in ohms, and C in farads.

The units may cause difficulty, but if the resistance is expressed in megohms (e.g. 50 kilohms = 0.05 megohms), C can be expressed in microfarads. The electrolytic capacitors used on the capacitor board are not particularly accurate, and wide variations may be experienced in practice.

Further Work

In its present form, the LED glows during the temporary state of the monostable. Pupils could modify the circuit to cause the LED to stop glowing when the switch is pressed, and glow again at the end of the timed period. A possible solution is given below.

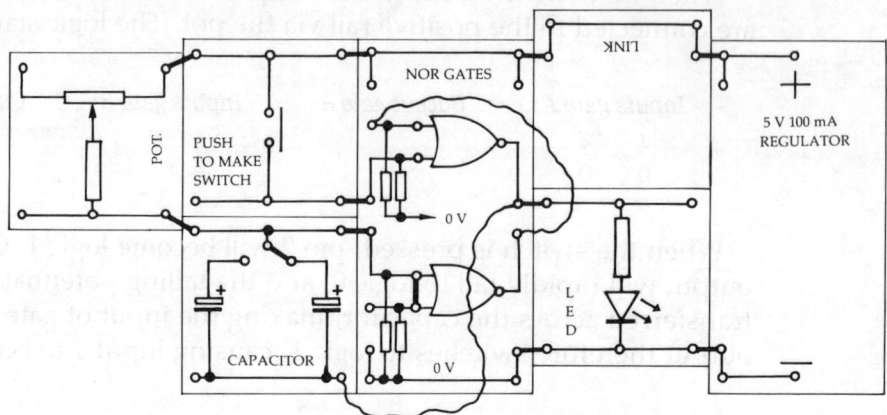

The LED could also be replaced with a transistor/relay/buzzer as used in previous circuits.

Homework
Suggestion

The house alarm system (Investigation 37) could be made to latch on, making the buzzer or siren sound for an indefinite period. This is both unwise (and possibly illegal) in practical systems, since a siren sounding indefinitely could cause annoyance to neighbours if, for example, the house occupants were away on holiday. Pupils could be asked to link the alarm circuit to the monostable so that the siren sounds for only 15 minutes. The solution is shown below.

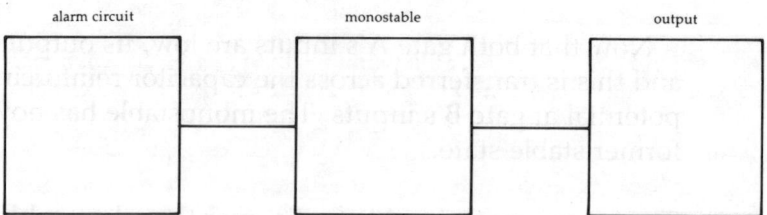

Note that the monostable circuit is placed between the alarm and the output transistor/relay. Pupils could use the Kent Electronics Kit to test their design, but the time period required would have to be reduced to 50 seconds, using the components available.

Background
Information

Specialised monostable ICs are available, such as the type 74HC221 which contains two monostables, or type 74HC123 which is similar except that it can be retriggered to increase the timed period. A very inexpensive and popular monostable is the NE555V IC. Capable of sourcing or sinking a current of up to 200 mA, this IC also boasts a much wider voltage range than 74HC CMOS.

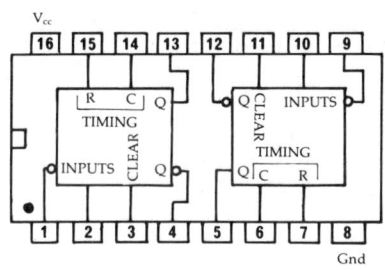

74HC221 Dual monostable multivibrator

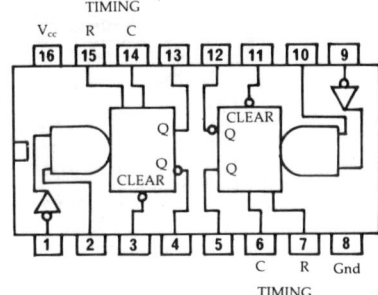

74HC123 Dual monostable multivibrator — retriggerable

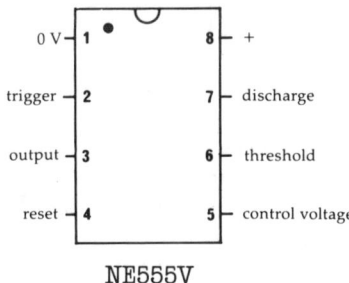

NE555V

Investigation 41
Astable Multivibrator

In Circuit 41 the LED should flash on and off at a speed determined by the setting of the pot. and the value of the capacitor.

Note: If the capacitor link is removed, or if the pot. is rotated fully clockwise, the LED will flash very quickly and appear 'half bright'. The circuit will then become unstable, and it is no longer possible to reduce the flashing rate, except by switching off, rotating the pot. to its midway position and switching on again.

Circuit Theory   The *astable* has no stable state, and constantly changes from high to low.

Both sets of inputs to the NAND gates are linked, making them behave as inverters. Assume that gate B's output is low (LED off). This will hold the inputs to gate A low. The voltage on the positive side of the capacitor will fall, as current flows via the pot. towards the output of gate B. Since this capacitor is connected to the inputs of gate B, when

129

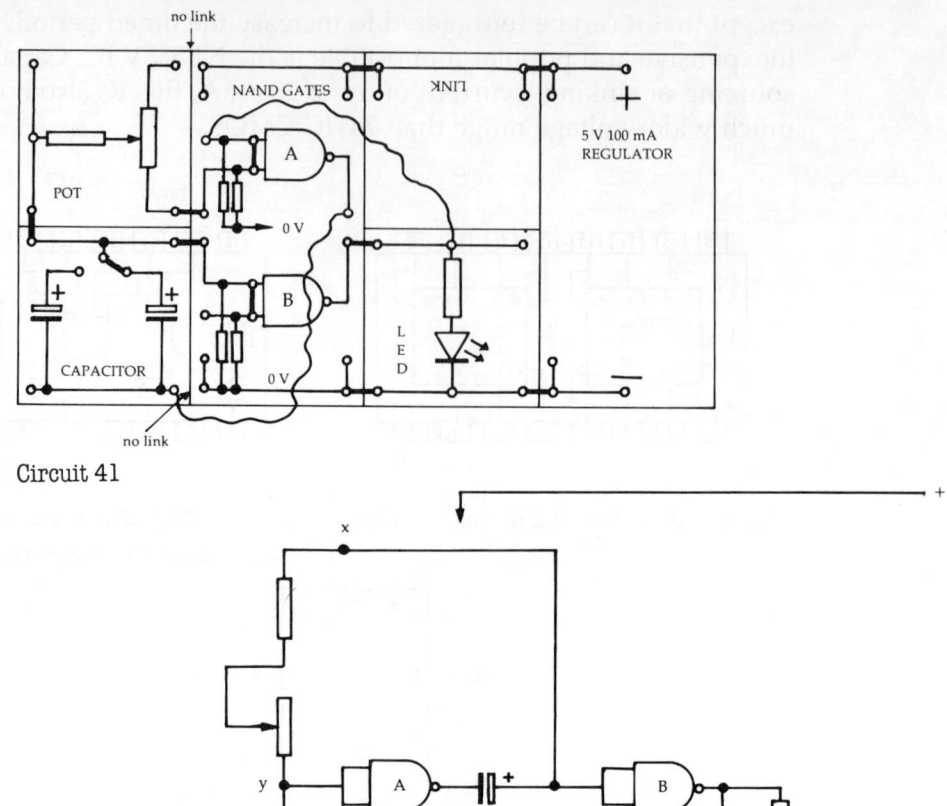

Circuit 41

this voltage reaches logic 0, gate B's output will go high, making the LED glow.

Gate A's inputs will go high, and current will flow via the pot. causing the voltage on the positive side of the capacitor to rise, eventually reaching logic 1. This will cause gate B to change state again, its output going low. The LED will now be turned off.

This process will continue to repeat itself, at a rate (frequency) determined by the time taken for the capacitor to charge and discharge. This depends upon the value of the capacitor and the setting of the pot.

Further Work This astable provides a square wave output, which is ideal for driving other logic devices, such as counters. The output may also be used to drive an earpiece or loudspeaker to produce a series of 'clicks'. The circuit may drive a high impedance earpiece connected in place of the LED; alternatively, an output transistor could be connected, with a 64 Ω (or more) loudspeaker (or an 8 Ω loudspeaker with a series resistor of about 50 Ω), as shown.

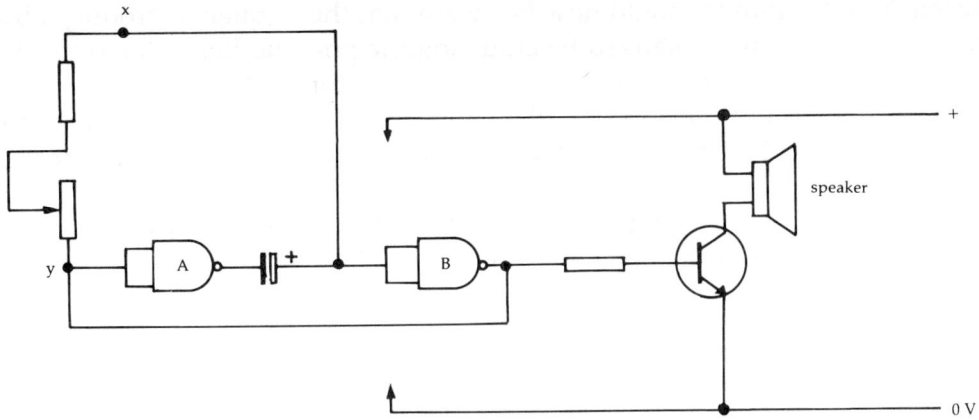

The circuit is now effectively a metronome. The output may be connected to an oscilloscope, though in order to obtain the best results it may be necessary to disconnect the loudspeaker (if still connected).

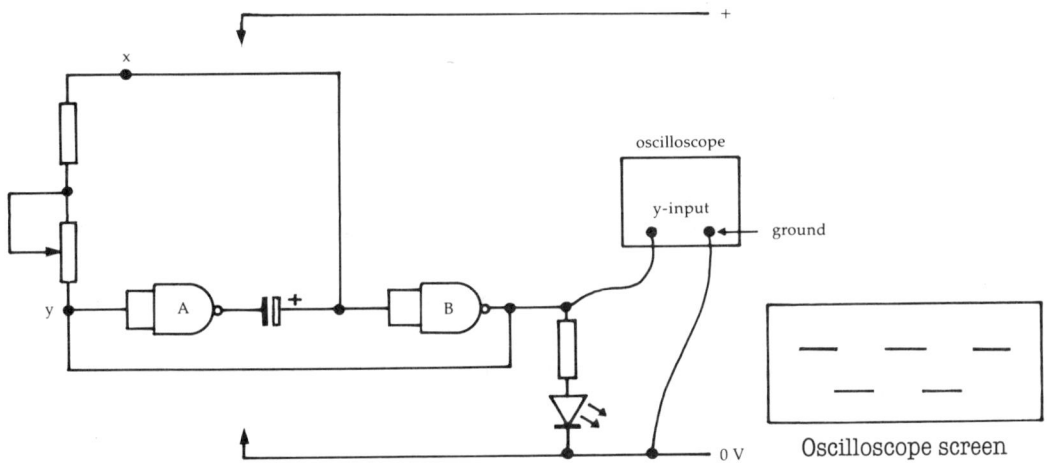

Oscilloscope screen

What the screen shows is, in fact, a square wave, but the output changes between logic levels so quickly that no vertical trace is left on the screen. Pupils may prefer to add the missing vertical lines, as shown opposite.

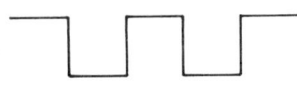

Increasing the Frequency

If the frequency is increased, a musical note is produced from the loudspeaker. This is achieved by using a 0.1 μF capacitor in place of the 22μF one in use at present. Small 0.1 μF ceramic or polyester capacitors can be purchased for a few pence, and could be connected using flying leads. A wide range of notes could then be produced from the loudspeaker.

Pupils should now be aware that the frequency produced by the circuit can be changed by adjusting the pot. and hence the resistance between the input to gate A and the capacitor. If an earpiece or speaker is connected as described above, and a small capacitor (e.g. 0.1 μF) is used, the circuit could become a monotone (i.e. one note at a time) electronic organ. Pupils could be asked to design a one-octave organ by using eight pots. and a 'probe' (a piece of wire pushed through an old plastic pen case may be used). The solution is given below.

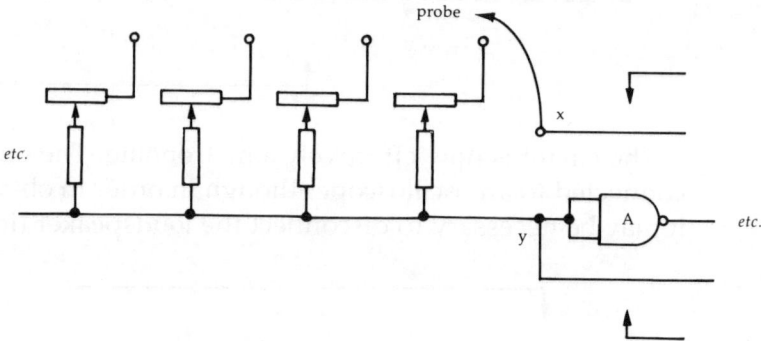

The circuit could be built, but a set of eight to ten pots. is required to make a usable system. The organ may be tuned by making the probe touch the appropriate contact and adjusting each pot. to produce a different note.

A teacher, technician or pupil with soldering experience could produce an inexpensive 'keyboard' using preset potentiometers (known as 'miniature presets' in electronics catalogues). These can be purchased for a few pence each, and a set of eight or more could be arranged as shown below.

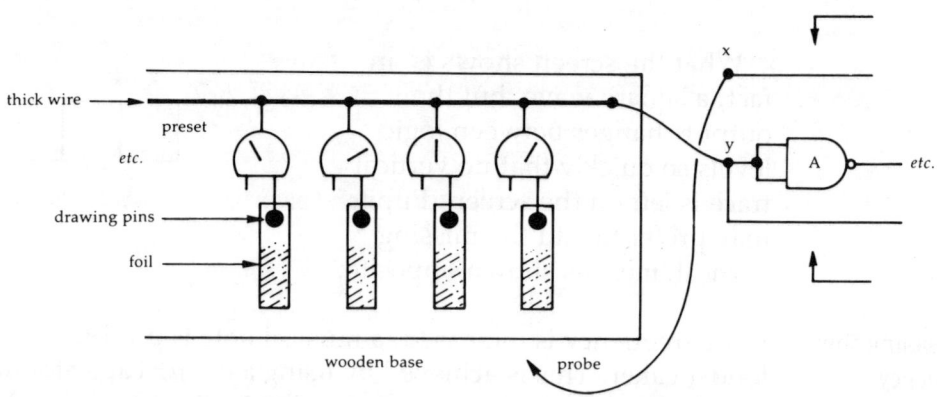

The 'keyboard' could be connected to the circuit via a flying lead, and played by using a probe, as before.

Investigation 42
Binary to 7-Segment Display

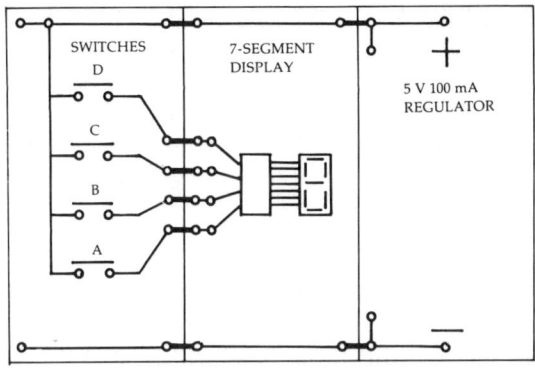

Circuit 42

74HC4511

6	D	a 13
		b 12
2	C	c 11
		d 10
1	B	e 9
		f 15
7	A	g 14

680R

switches board   7-segment display board

100 K   1K5

+

0 V

When Circuit 42 is switched on, a 'zero' should be displayed. When switch A is pressed, a number '1' should appear, switch B should produce a '2', C a '4', and D an '8'. Combinations of switches should produce the other numbers. The solution for the first table in the Pupil's Book is given overleaf.

133

Switches	Number displayed
None	0
A	1
B	2
B and A	3
C	4
C and A	5
C and B	6
C and B and A	7
D	8
D and A	9

If an attempt is made to produce a number over 9, the display goes blank.

The solution for the second table is as follows:

Binary number				Decimal number
D	C	B	A	
0	1	0	1	5
0	0	1	1	3
1	0	0	1	9
1	0	0	0	8

The above table may be tested, noting that a switch pressed represents logic 1, and not pressed, logic 0.

Homework Suggestion

Conversions between binary and decimal numbers could be set to reinforce the ideas learned in this investigation.

Background Information

The 7-segment display was discussed on pages 45 to 47. Each segment is an LED, each cathode of which is connected to zero volts via a single pin. This type of device is known as a *common cathode* display. The seven anodes are connected via current-limiting resistors to the IC outputs.

The IC used is a type 74HC4511. This is known as a 'BCD to 7-segment latch/decoder/driver'. In other words, it converts a binary coded decimal input to a 7-way output, capable of producing decimal numbers on the display.

The LEDs indicate when an input is high. Each LED requires a current-limiting resistor, and the rather high value of 1.5 kΩ is chosen to limit the total current consumed by the display board. Each input is held low via a 100 kΩ resistor. While the LEDs tend to hold the inputs low anyway, the forward voltage drop across the LED could allow unwanted voltages at the data inputs. Each output can supply up to 10 mA, and is connected to the appropriate LED segment via a current-limiting resistor.

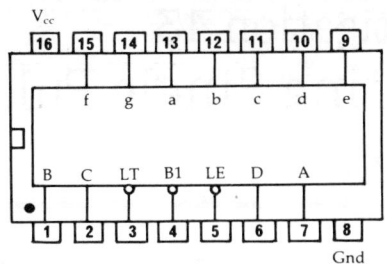

74HC4511 BCD-7 segment latch/decoder/driver

The diagram above shows the pinout arrangement of the 74HC4511. Pins 1, 2, 6, 7 are the four binary data inputs. Pin 3 is a 'lamp test'. When pin 3 is high the IC functions normally, but when low, all segments of the display light up. Pin 4 is a 'ripple blanking input'. When pin 4 is high the IC functions normally, but when low, no segments light up, regardless of the data inputs. Pin 5 is a 'latch enable'. When pin 5 is low the IC functions normally, but when high, the last data present at the data inputs is stored in the latches, causing the display to remain stable regardless of changes at the data inputs. Since the functions provided by pins 3, 4 and 5 are not required, pins 3 and 4 are connected to positive, and pin 5 to zero volts.

The output from the IC is fed via pins 9 to 15, each of these being connected (via a resistor) to a segment, 'a' to 'g' on the display as shown below.

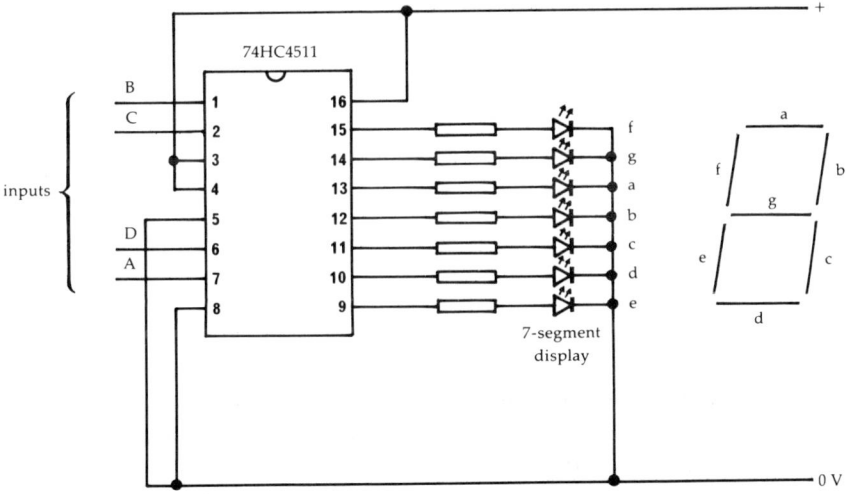

Investigation 43
Counting Single Pulses Using a Push Switch

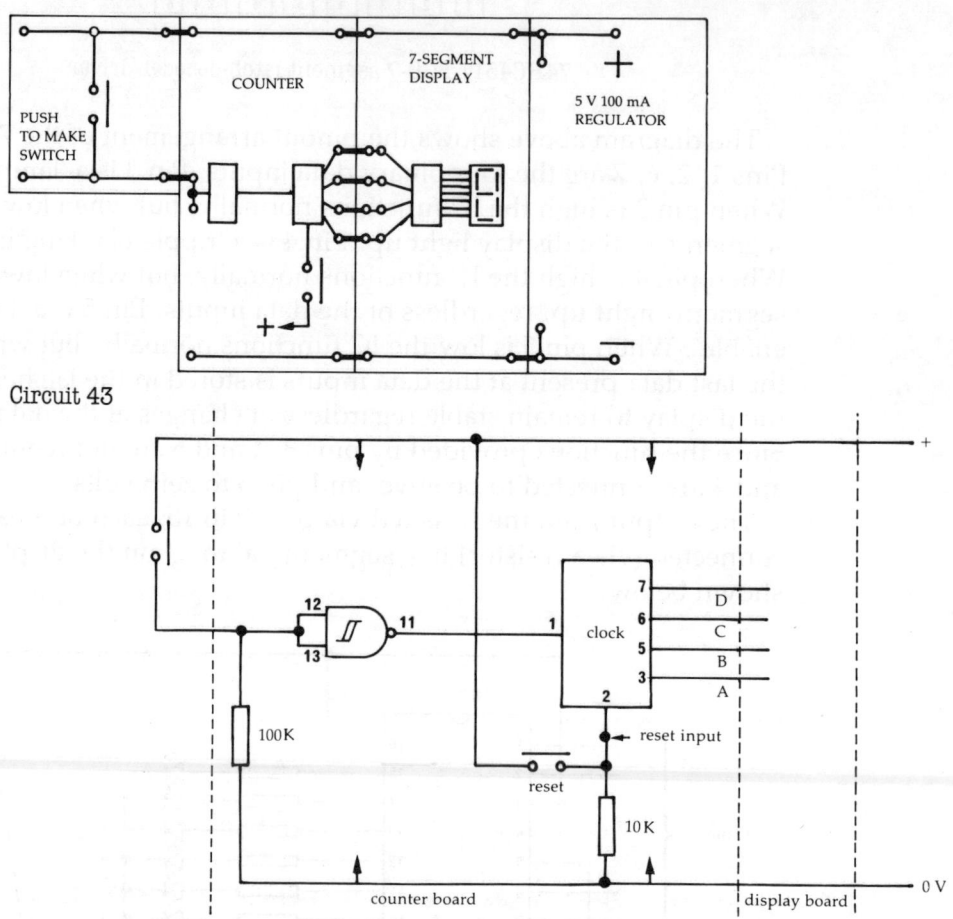

Circuit 43

When Circuit 43 is switched on, a number (which is not always zero) is displayed. The 'reset' switch may be pressed to return the counter to zero. Each time the other push switch is pressed the display should increment by one. However, it sometimes increments by more than one as a result of switch contact 'bounce'. Bounce occurs because the switch contacts do not close or open smoothly. As the switch is pressed, the contacts may close then re-open several times before closing properly. This happens in a fraction of a second, but the pulses produced cause the counter to increment incorrectly.

Some push switches function better than others in this respect. The push switches supplied with the Kent Electronics Kit should produce quite good results if operated 'cleanly', i.e. with sharp presses. The reed switch may produce more contact bounce, to such an extent that normal counting is impossible.

Clearly, switch contact bounce is a serious problem in digital electronics, and this investigation illustrates this, in addition to showing how a binary counter may be employed.

Background
Information

The binary counter board houses two ICs, the left-hand IC being a Schmitt trigger type 74HC132. This is known as a quadruple 2-input NAND Schmitt gate. Only one gate is required, all the other inputs being connected to zero volts. The two inputs of the Schmitt NAND gate in use are connected together, making the gate a Schmitt inverter, i.e. a Schmitt trigger connected to an inverter. A Schmitt trigger provides a very 'clean' output which is either fully at logic 0 or logic 1, even if a slowly rising or falling voltage is applied at the inputs. (The Schmitt trigger was discussed on page 101.) The inverter is required because the counter IC counts each time it receives a *negative* pulse.

To sum up, a slowly rising positive voltage applied at the input to the Schmitt inverter is converted to a very fast voltage drop from positive to zero, as required by the counter IC. The counter IC used is a type 74HC390, shown opposite. This is actually two 'decade counters' in one package.

The diagram below shows the complete circuit arrangement of the counter board. A pinout diagram of the 74HC132 IC is shown on page 102.

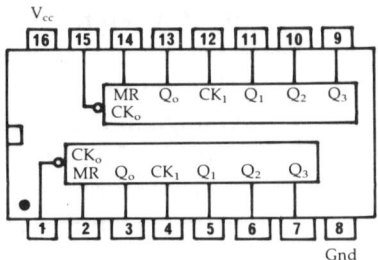

74HC390 Dual decade counter

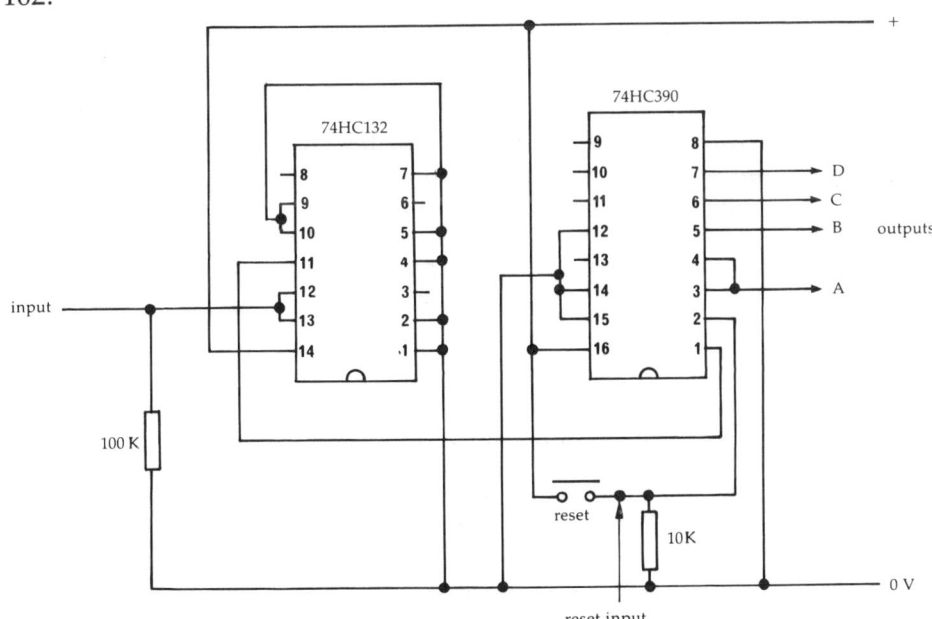

137

All the unused inputs (pins 1, 2, 4, 5, 9 and 10) are connected to zero volts. The counter board input is fed to pins 12 and 13, a 100 kΩ resistor being used to keep the input at zero volts unless a positive pulse is applied.

Although the 74HC390 IC includes two decade counters, only one is required. The unused inputs (pins 12, 14, 15) are therefore connected to zero volts.

The output from pin 11 of the Schmitt inverter is fed to pin 1 (known as the clock 0 input) on the counter IC. Various counting arrangements are possible, but for normal 'binary coded decimal' counting, pins 3 and 4 are connected together. The binary data outputs are via pins 3, 5, 6 and 7.

The counter IC may be reset by applying a positive voltage to pin 2. The 10 kΩ resistor holds the reset low unless the push switch is pressed, or a positive voltage is applied at the reset socket.

Binary Coded Decimal (BCD) Counting

The outputs function according to the rules outlined in Investigation 42. BCD counting means that the counter automatically resets to zero when it counts beyond 9. A standard binary counter with four outputs will reset when it counts beyond 15.

Resetting

The counter can be forced to reset at any time by applying a positive voltage at the reset socket. For example, a flying lead could be used to connect data line C to the reset socket. The counter will now reset each time it counts beyond 3. This function is particularly useful when further counters are cascaded, as described in Investigation 47.

Further Work

A wide variety of circuits may be connected to the counter input. Some particular examples are outlined in the remaining investigations. Pupils may experiment by connecting any of the preceding circuits to the counter.

Homework Suggestion

Pupils could be asked to design a circuit which counts the number of times a door opens. At this stage they should disregard the problem of switch contact bounce, and the limitation imposed by a maximum count of 9.

Investigation 44
Anti-Bounce Circuit

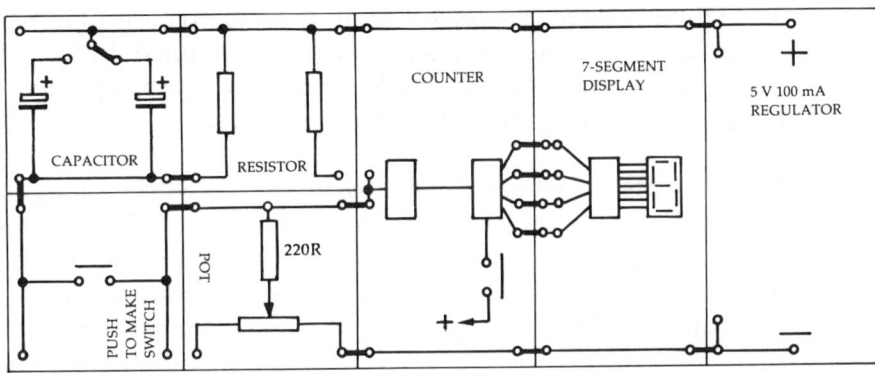

Circuit 44

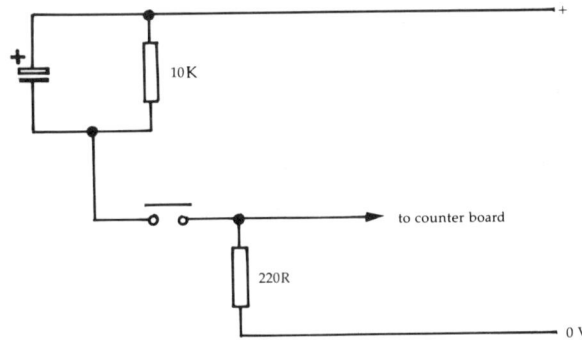

It is now clear that switch contact bounce is a considerable problem in digital electronics. There are many ways of reducing or eliminating the problem. This circuit should reduce the problem sufficiently to allow quite poor switches to be employed. However, the components are chosen because of their availability in the Kit, and are not ideal.

It is important to rotate the pot. fully anti-clockwise, so that its resistance is zero. The 220 Ω resistor then becomes the only effective resistance on the pot. board.

When the push switch is pressed, the counter should increment by one, without the 'jumping' associated with contact bounce. The reed switch should also give much better results.

Circuit Theory

When switched on, the 10 kΩ resistor causes the lower side of the capacitor to be at 5 V. The upper side is also at 5 V, thus the capacitor is *discharged*. *Remember*: A capacitor is discharged if *both* sides are at the same potential.

The input to the counter is held at zero volts by the 220 Ω resistor.

The resistor already fitted on the counter board has a much higher value and will be disregarded.

When the switch is pressed, a very brief positive pulse is applied to the counter input. The lower side of the capacitor rapidly falls to zero volts via the 220 Ω resistor. The capacitor is now said to be charged. The fall in voltage is so fast that, in theory, no current is available to cause false counting if the switch contacts bounce. In this circuit the capacitor is larger than desirable, and some bouncing may still be experienced with particularly poor switches.

Further Work

If the counter functions reliably with a reed switch, a magnet could be fixed to a pendulum to enable automatic counting of 'half swings'. The magnet could also be fitted to a rotating shaft to determine its speed. The number of zeros which appear on the display should be counted, and multiplied by ten to produce the total number of revolutions in a given time. Care must be taken to ensure that the magnet does not fly off into a pupil's eye.

Investigation 45
Counting Using a Light Beam

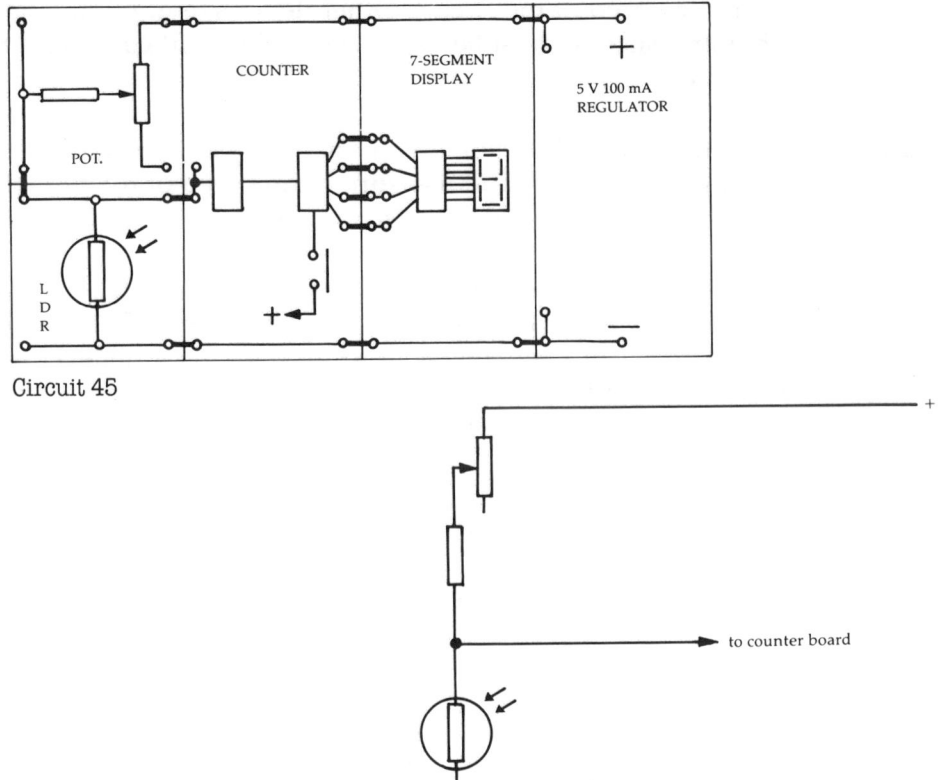

Circuit 45

The setting of the pot. is critical in this circuit, and pupils should be advised to first check that the display will increment when the pot. is turned from fully clockwise to fully anti-clockwise.

Next, repeat this operation very slowly, stopping when the counter increments. Turn the pot. back a little, and see if the counter increments when the LDR is shaded.

The circuit should work in an ordinary classroom situation, with a mixture of artificial lighting and daylight. The sensitivity can be increased by using more direct light, for example by using daylight from the windows only, or – for best results – by using a beam of light from a projector or ray box. The room lighting could then be subdued, or a hood could be fitted around the LDR. The circuit should then be able to count objects passing through the beam of light.

Background Information

The importance of the Schmitt trigger (*see* page 137) fitted on the counter board should not be underestimated. Circuit 45 produces a slowly changing voltage as the LDR is shaded, and will not operate the clock input of the counter IC directly. The Schmitt trigger converts the slowly changing voltage into a very sudden change.

The counter IC is capable of operating at high speed, but the LDR does not react very quickly to changes in light level and will limit the counting speed of the circuit. Other devices (e.g. photodiodes and phototransistors) must be used (with a different circuit) for high-speed counting.

Applications

Counting objects passing on an conveyer belt, counting people passing through a doorway, and measuring the speed of rotation of a device such as a fan are just three of many applications for this circuit.

Further
Work/
Homework

Any of the binary outputs A to D may be connected via a flying lead to the base of a transistor, or the input of another logic gate. Pupils could devise an alarm system which sounds a buzzer whenever the counter reaches a count of say, 8. Data line D would be connected to a transistor/relay/buzzer circuit as shown below.

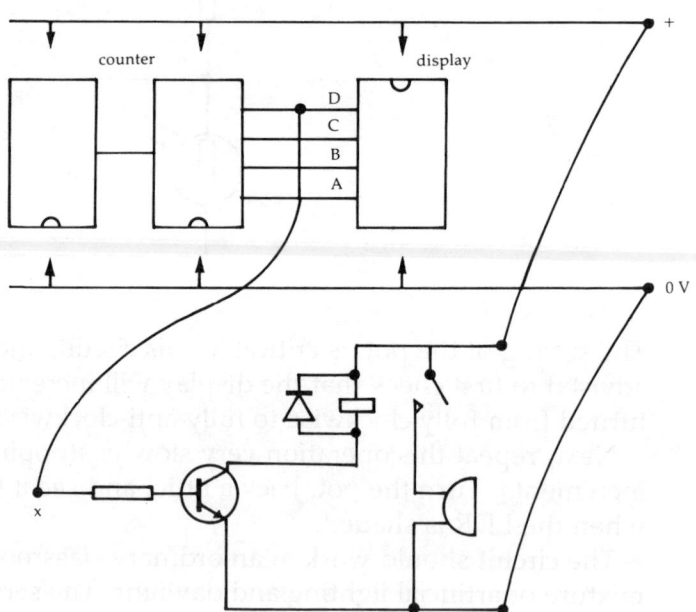

A NAND gate followed by an inverter, to create an AND function, could be used to sound the alarm at other numbers. If data lines B and C are fed to an AND gate, as shown below, the output will go high each time the counter reaches 6.

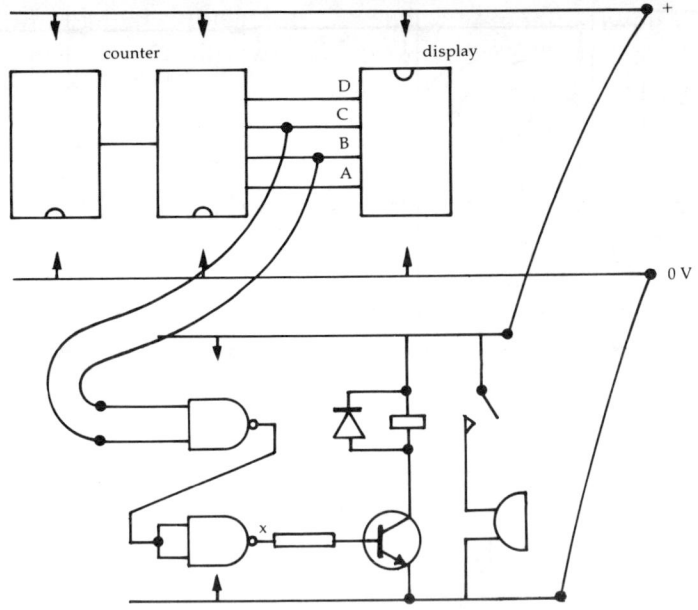

This type of work could be extended when further counters are cascaded, as shown in Investigation 47.

Investigation 46
Astable Multivibrator and Counter

When using this circuit, pupils should begin by checking that the LED flashes on and off correctly. Remember that the circuit may become unstable if the pot. is turned fully clockwise or a capacitor link is removed, and this can only be cured by switching off the supply for a moment. Each time the LED glows, the display should increment.

The astable circuit and counter circuit have already been explained, and this investigation merely links the two together.

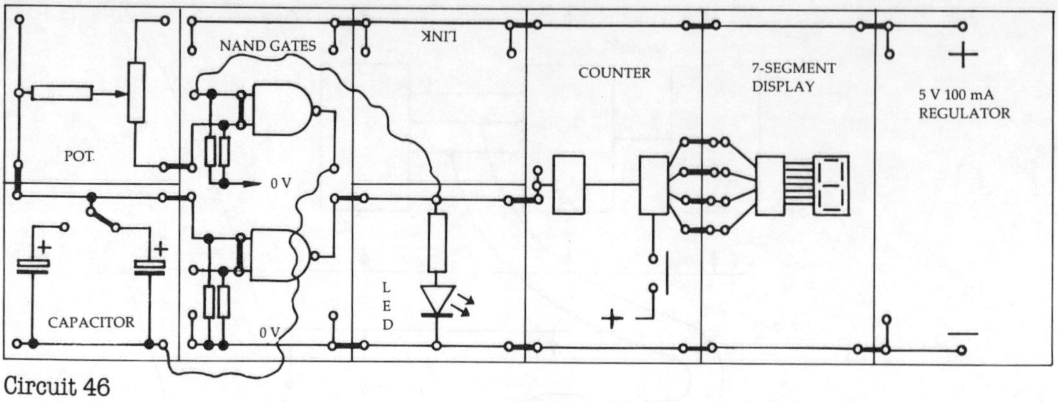

Circuit 46

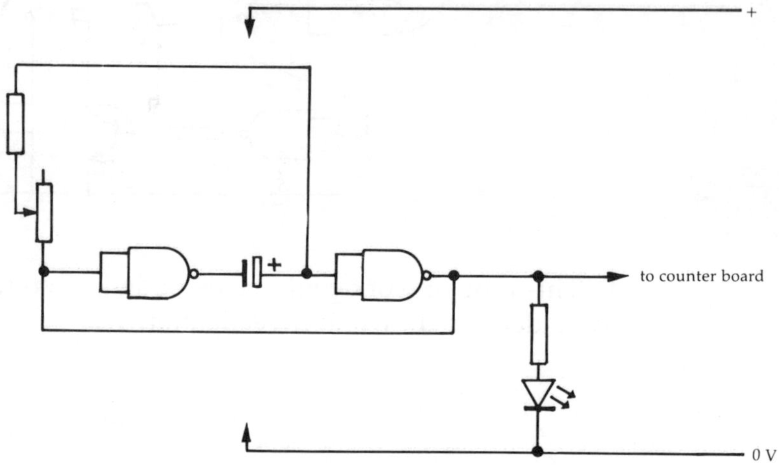

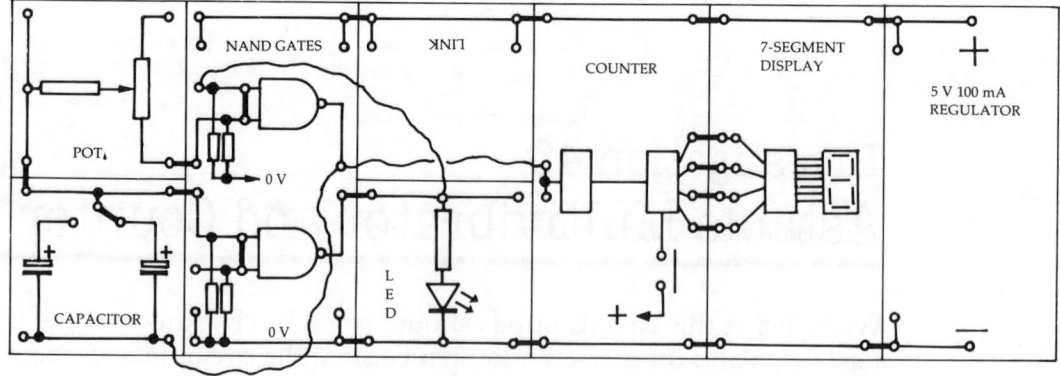

to counter board

0 V

If the link between the counter and LED is removed, as in the diagram below, and a flying lead added as shown, the display will increment when the LED switches *off*.

On some occasions the display may increment at the wrong times. This is often caused by interference on the supply rails or elsewhere. The problem may be reduced by connecting a 10 kΩ resistor between the output of the NAND gate and the input of the counter.

Pupils could be asked to discover how the counter may be started and stopped using a push switch. One solution is to remove the link connecting the lower gate inputs together, so that it functions as a NAND gate rather than as an inverter. The spare input may then be used to start and stop the counter.

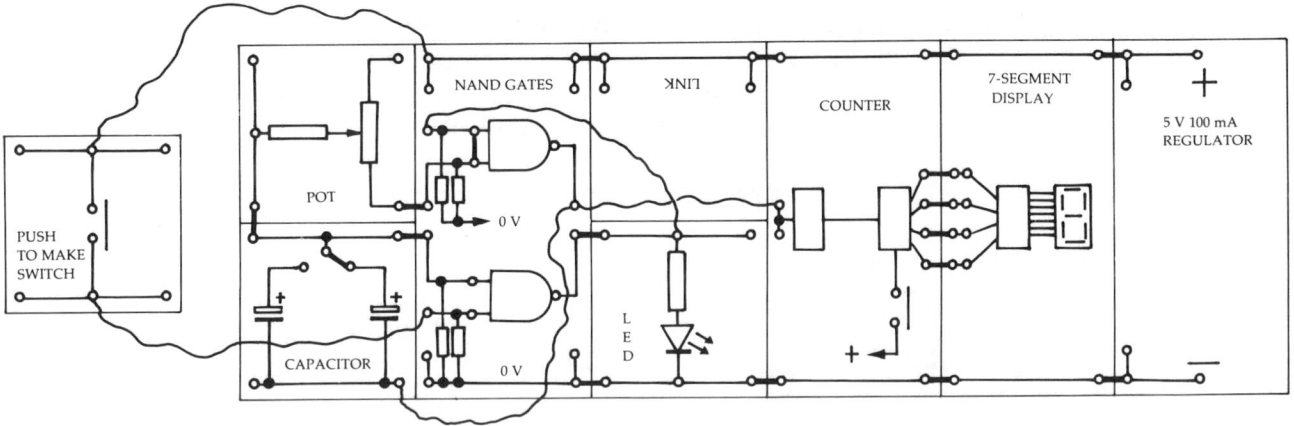

If the pot. is adjusted so that seconds are counted, the circuit could be used as a stopwatch.

The NAND gates board could be replaced with the NOR gates board. The NAND gates may then be used to make a bistable multivibrator with two push switches (as shown on page 124). If the output from the bistable is connected to the spare input on the lower NOR gate, this

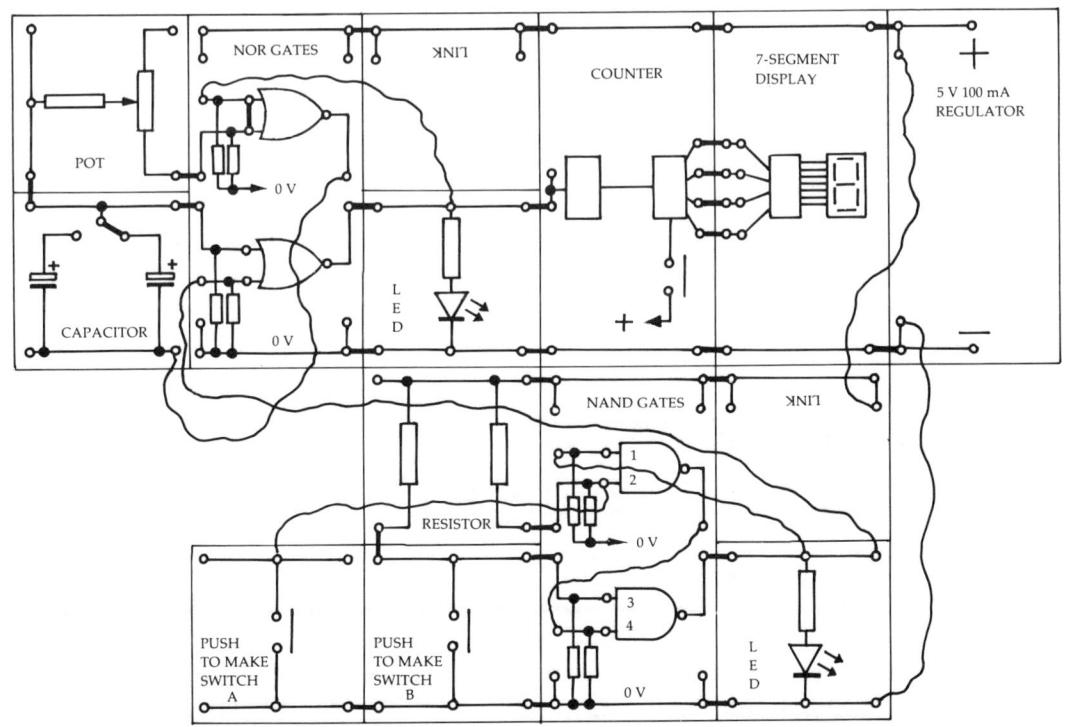

circuit will also function as a stopwatch, with a start button and stop button.

The above circuit counts when the LED is off, and stops counting when the LED glows. This may seem illogical, and pupils could suggest a modification so that the counting occurs when the LED glows. The simplest alteration would be to use the output from the upper gate of the bistable to control the astable and counter.

The circuit could also be started and stopped automatically by means of the LDR and another pot. The LDR and additional pot. should be connected to the spare input as shown below.

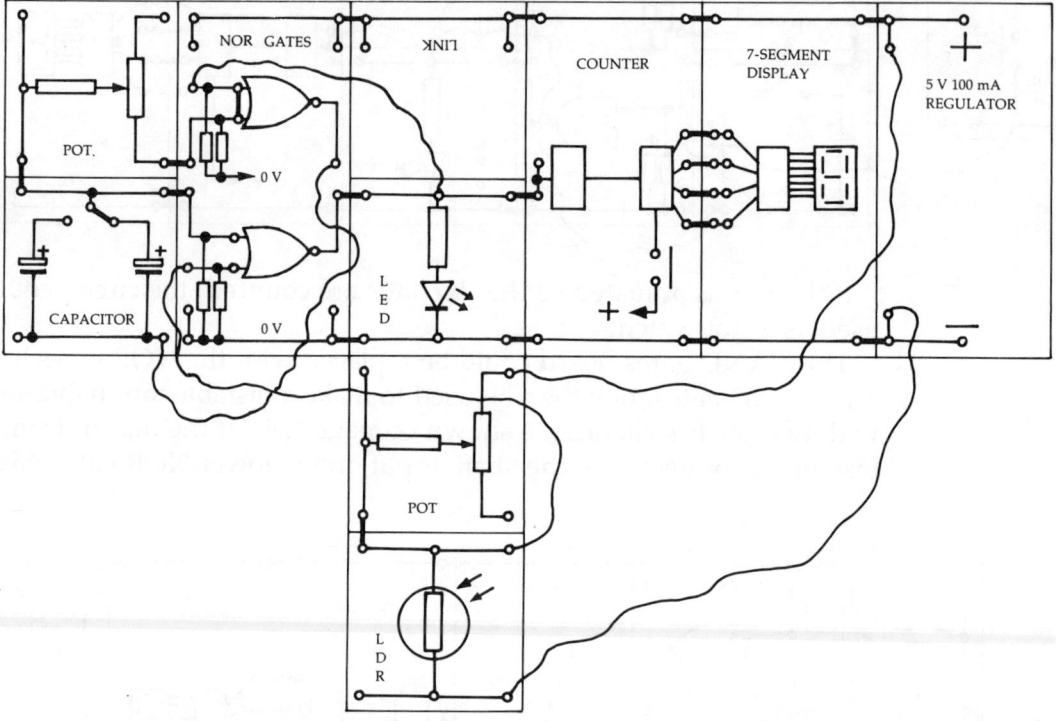

This circuit may be unstable in practice, since a slowly changing level of light produces an uncertain logic level. A Schmitt trigger is required between the junction of the LDR and pot., and the gate input.

Investigation 47
Astable Driving a Units and Tens Counter

In Circuit 47 the upper circuit should work as before; the lower counter circuit should increment each time the upper display counts beyond 9.

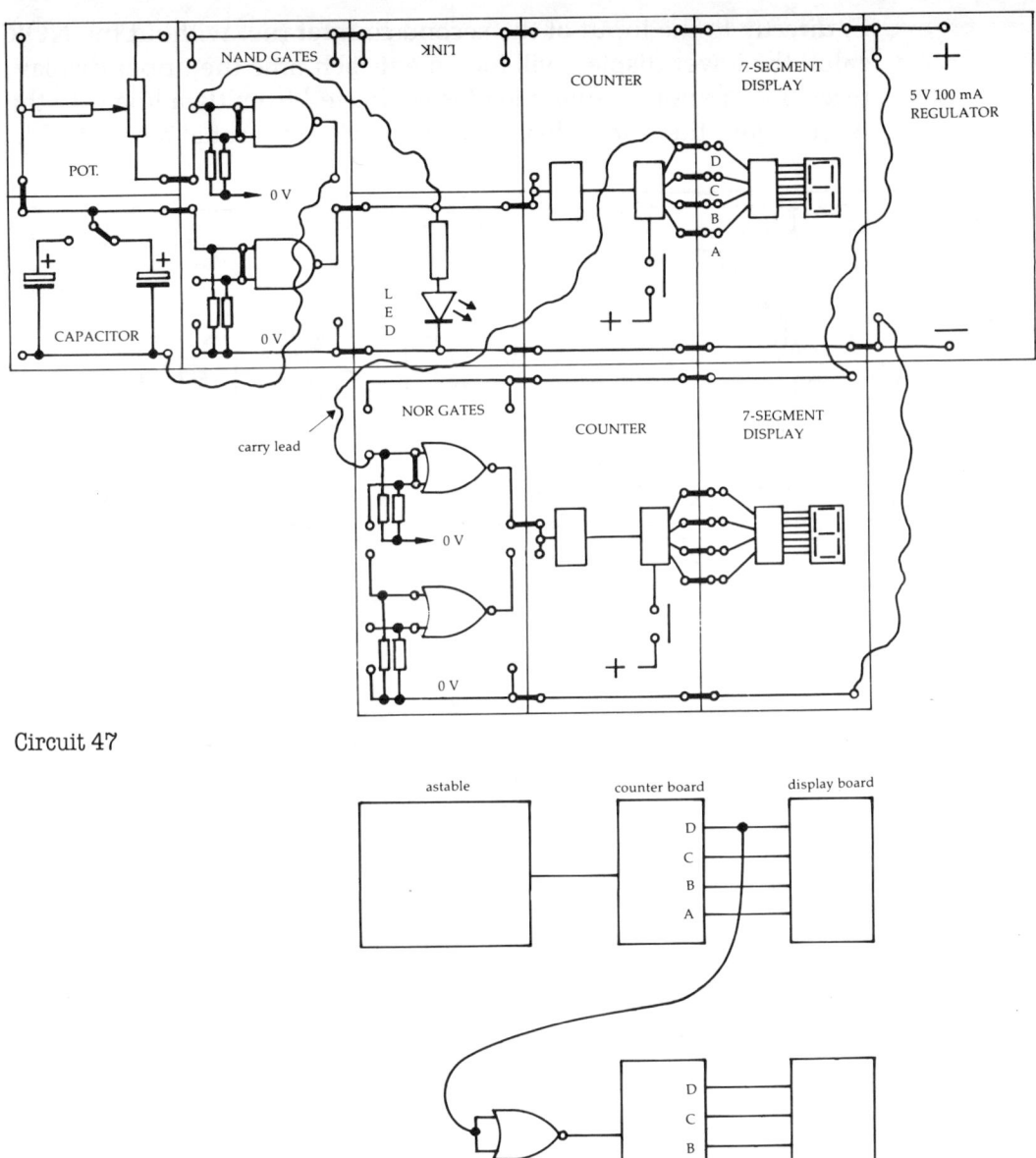

Circuit 47

The binary data line D changes from high to low each time the counter counts beyond 9. However, the second counter board requires a *rising* voltage to cause the counter to increment. A NOR gate is therefore required, connected as an inverter to produce a rising voltage each time its input changes from logic 1 to logic 0.

All the previous counter circuits may be extended in this way, to enable counting to 99.

Further Work   Some interesting experiments can be performed with a variety of number bases. For example, if the carry lead is connected from data line

147

C, directly to the input of the second counter (i.e. without the NOR gate), the lower display will increment each time the upper display reaches 4. If another flying lead is connected from data line C to the reset socket, the upper display will 'carry' and reset at a count of 4.

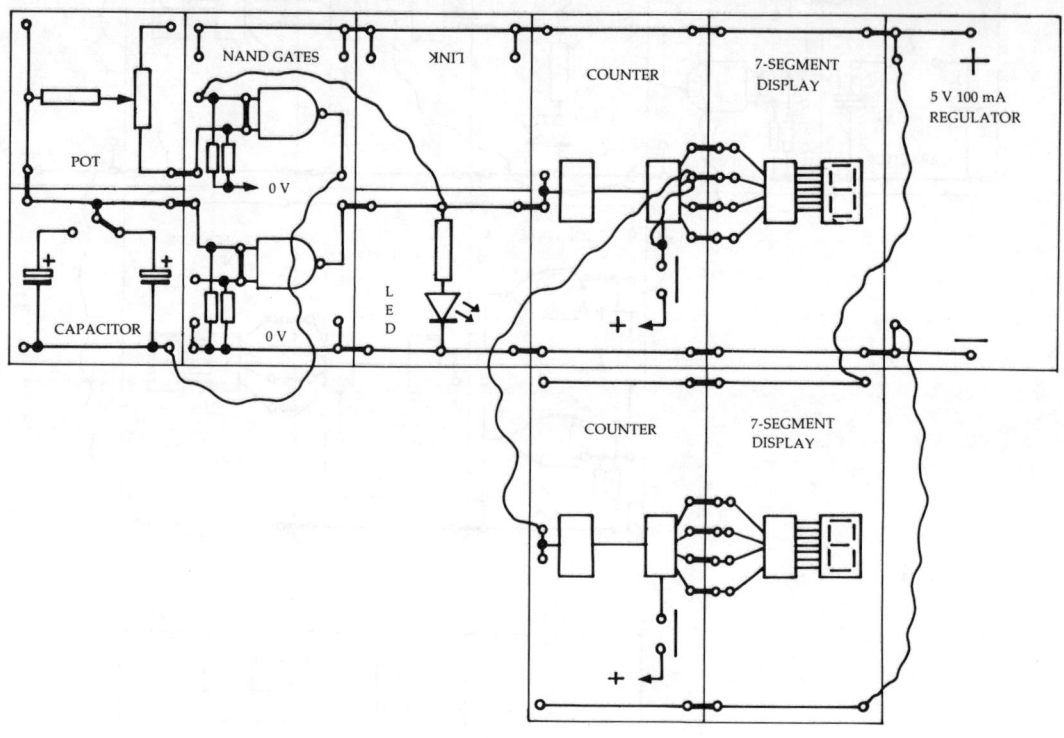

An AND gate (i.e. two NAND gates in series) could be used to combine data lines, as shown on the page opposite, for example to make the upper display 'carry' and reset on a count of 6.

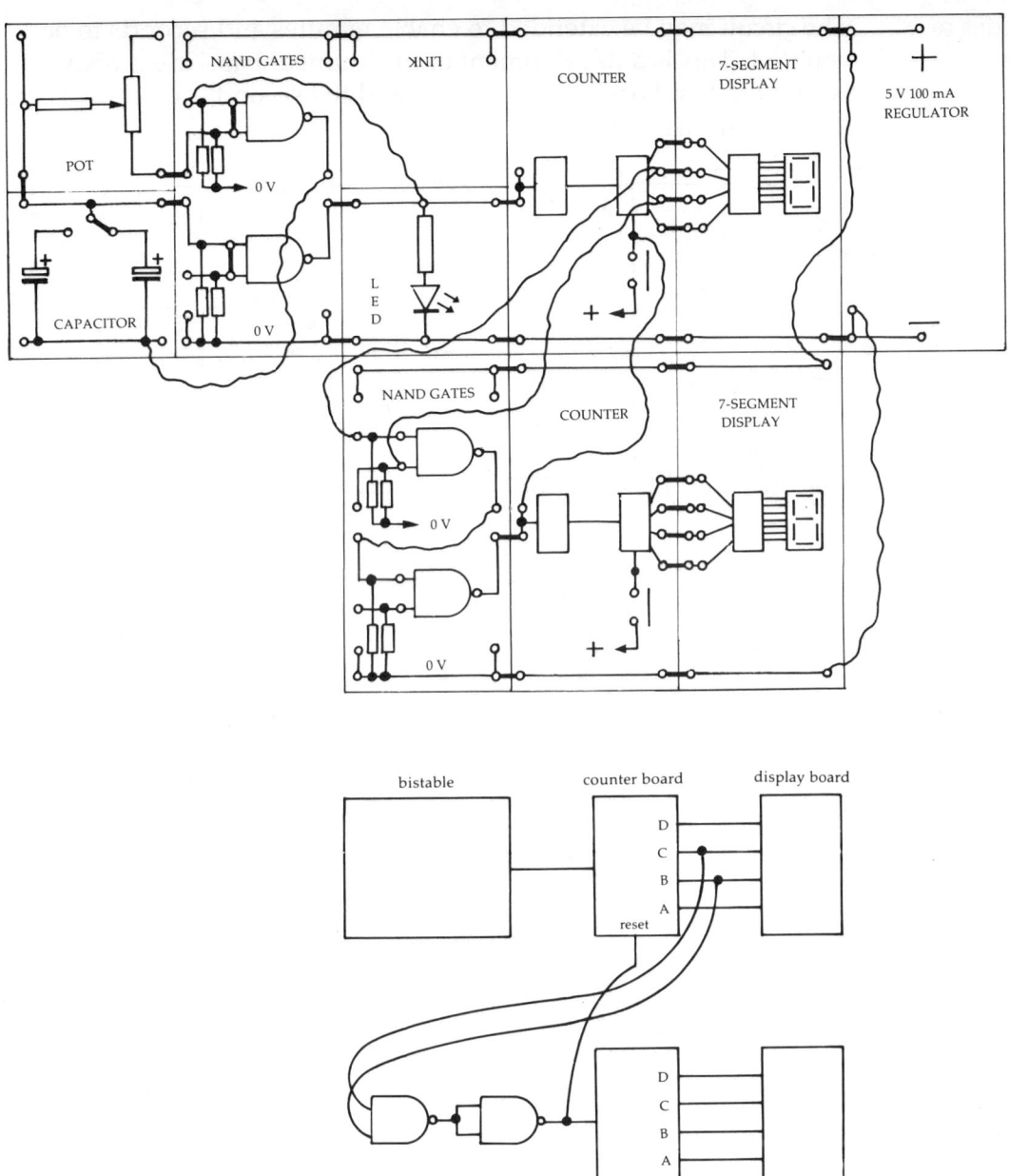

Adding a Third Display

Another set of boards may be added to enable counting to 999. In trials it was found that the regulator was capable of supplying all three sets, though it was not designed for this purpose. A second regulator may be connected if desired, though care must be taken to switch on both regulators at the same time (ideally they should be powered by the same power unit). The negative (0 V) rails of the regulators should be connected together. (If they are powered from the same power unit, the negative sides will already be connected.)

The circuit may be extended to enable minutes and seconds to be counted. This is a development of the 'number base' ideas above, where the first board 'carries' at 10, and the second 'carries' at 6. The circuit is shown below.

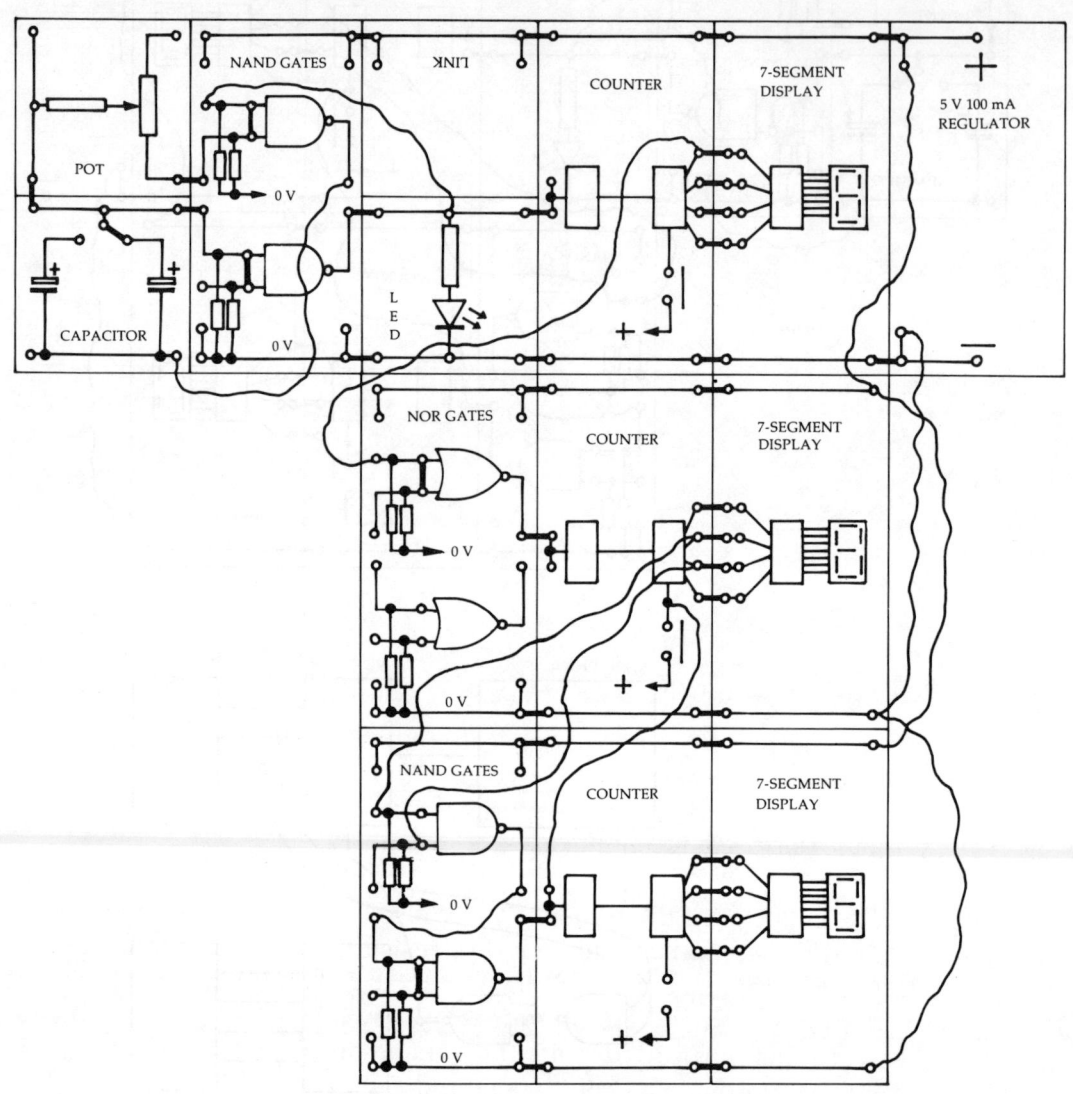

150

Investigation 48
Connecting a Logic Gate and LED

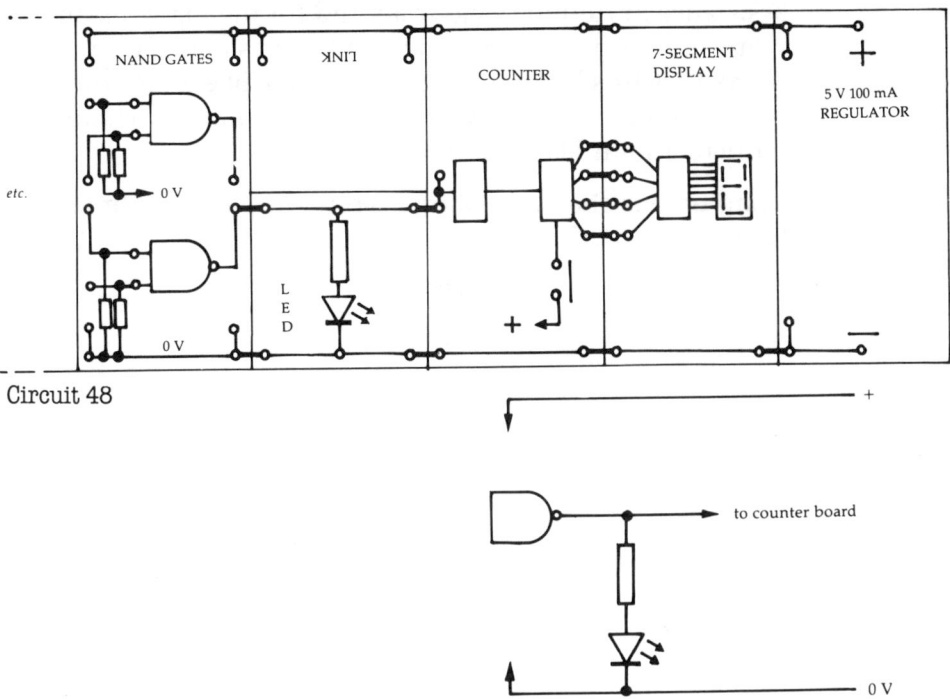

Circuit 48

The method by which a logic gate output can be connected to the counter has already been covered, but Circuit 48 is included to suggest an open-ended series of experiments, where any circuit employing a logic gate may be connected to the counter.

Further Work

The collector of the transistor may also be connected to the counter input. It is important to include a resistor between the collector and the positive rail, so that the voltage at the collector is nearly at the positive supply voltage when the transistor is switched off. The LED board will serve this purpose (despite the voltage drop imposed by the LED) and will also provide a useful visual indication.

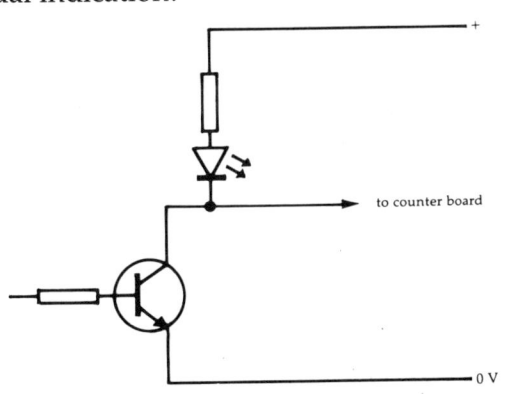

151

ACKNOWLEDGEMENTS

The authors would like to thank Paul Horton (Cornwallis School), Melvin Chase and Roy Crowhurst (both Clare Park School) for their assistance in running field trials and for the many useful suggestions, many of which have been included in this book. They would also like to thank Dr Chris Bounds (Christ Church College) for proofreading the text, and the pupils of Warren Wood Secondary School for Boys, in particular John Blaker and Ross Banyard.

Components shown in photographs supplied by Omni Electronics, Dalkeith Road, Edinburgh.